Communications in Computer and Information Science 584

Commenced Publication in 2007
Founding and Former Series Editors:
Alfredo Cuzzocrea, Dominik Ślęzak, and Xiaokang Yang

More information about this series at http://www.springer.com/series/7899

Markus Helfert · Andreas Holzinger
Orlando Belo · Chiara Francalanci (Eds.)

Data Management Technologies and Applications

4th International Conference, DATA 2015
Colmar, France, July 20–22, 2015
Revised Selected Papers

 Springer

Editors

Markus Helfert
School of Computing
Dublin City University
Dublin
Ireland

Andreas Holzinger
Human-Computer Interaction
Medical University of Graz
Graz
Austria

Orlando Belo
Department of Informatics
University of Minho
Braga
Portugal

Chiara Francalanci
Department of Electronics and Information
Politecnico di Milano
Milan
Italy

ISSN 1865-0929   ISSN 1865-0937 (electronic)
Communications in Computer and Information Science
ISBN 978-3-319-30161-7   ISBN 978-3-319-30162-4 (eBook)
DOI 10.1007/978-3-319-30162-4

Library of Congress Control Number: 2015952775

Printed on acid-free paper

This Springer imprint is published by SpringerNature
The registered company is Springer International Publishing AG Switzerland

Preface

The present book includes extended and revised versions of a set of selected papers from the 4th International Conference on Data Technologies and Applications — DATA 2015), which is sponsored by the Institute for Systems and Technologies of Information, Control and Communication (INSTICC), and co-organized by the University of Haute Alsace and held in cooperation with the ACM SIGMIS — ACM Special Interest Group on Management Information Systems.

The aim of this conference series is to bring together researchers and practitioners interested in databases, data warehousing, data mining, data management, data security, and other aspects of knowledge and information systems and technologies involving advanced applications of data.

DATA 2015 received 70 paper submissions, including special sessions, from 32 countries in all continents, of which 44 % were orally presented (20 % as full papers). In order to evaluate each submission, a double-blind review was performed by the Program Committee.

The high quality of the DATA 2015 program was enhanced by the three keynote lectures, delivered by distinguished speakers who are renowned experts in their fields: Michele Scbag (Laboratoire de Recherche en Informatique, CNRS, France), John Domingue (The Open University, UK), and Paul Longley (University College London, UK).

The quality of the conference and the papers herewith presented stems directly from the dedicated effort of the Steering Committee, Program Committees, and the INSTICC team responsible for handling all logistical details. We are further indebted to the conference keynote speakers, who presented their valuable insights and visions regarding areas of hot interest to the conference. Finally, we would like to thank all authors and attendees for their contribution to the conference and the scientific community.

We hope that you will find these papers interesting and consider them a helpful reference in the future when addressing any of the aforementioned research areas.

July 2015

Markus Helfert
Andreas Holzinger
Orlando Belo
Chiara Francalanci

Organization

Conference Co-chairs

Markus Helfert	Dublin City University, Ireland
Andreas Holzinger	Medical University Graz, Austria

Program Co-chairs

Chiara Francalanci	Politecnico di Milano, Italy
Orlando Belo	University of Minho, Portugal

Program Committee

Muhammad Abulaish	Jamia Millia Islamia, India
Hamideh Afsarmanesh	University of Amsterdam, The Netherlands
Kenneth Anderson	University of Colorado, USA
Keijiro Araki	Kyushu University, Japan
Farhad Arbab	CWI, The Netherlands
Colin Atkinson	University of Mannheim, Germany
Orlando Belo	University of Minho, Portugal
Sukriti Bhattacharya	University College London, UK
Francesco Buccafurri	University of Reggio Calabria, Italy
Dumitru Burdescu	University of Craiova, Romania
Cinzia Cappiello	Politecnico di Milano, Italy
Krzysztof Cetnarowicz	AGH, University of Science and Technology, Poland
Kung Chen	National Chengchi University, Taiwan
Yangjun Chen	University of Winnipeg, Canada
Chia-Chu Chiang	University of Arkansas at Little Rock, USA
Stefan Conrad	Heinrich Heine University Düsseldorf, Germany
Agostino Cortesi	Università Ca' Foscari di Venezia, Italy
Theodore Dalamagas	Athena Research and Innovation Center, Greece
Bruno Defude	Institut Mines Telecom, France
Steven Demurjian	University of Connecticut, USA
Stefan Dessloch	Kaiserslautern University of Technology, Germany
Dejing Dou	University of Oregon, USA
Fabien Duchateau	Université Claude Bernard Lyon 1/LIRIS, France
Todd Eavis	Concordia University, Canada
Tapio Elomaa	Tampere University of Technology, Finland
Mohamed Y. Eltabakh	Worcester Polytechnic Institute, USA
Sergio Firmenich	Universidad Nacional de La Plata, Argentina
Chiara Francalanci	Politecnico di Milano, Italy

Nikolaos Georgantas   Inria, France
Paola Giannini   University of Piemonte Orientale, Italy
J. Paul Gibson   Mines-Telecom - Telecom SudParis, France
Boris Glavic   Illinois Institute of Technology Chicago, USA
Matteo Golfarelli   University of Bologna, Italy
Cesar Gonzalez-Perez   Institute of Heritage Sciences (Incipit), Spanish
                              National Research Council (CSIC), Spain
Janis Grabis   Riga Technical University, Latvia
Jerzy Grzymala-Busse   University of Kansas, USA
Aziz Guergachi   Ryerson University, Canada
Raju Halder   Indian Institute of Technology Patna, India
Barbara Hammer   Bielefeld University, Germany
Jose Luis Arciniegas   Universidad del Cauca, Colombia
  Herrera
Jose R. Hilera   University of Alcala, Spain
Andreas Holzinger   Medical University Graz, Austria
Jang-eui Hong   Chungbuk National University, Korea, Republic of
Tsan-sheng Hsu   Institute of Information Science, Academia Sinica, Taiwan
Xiaoxia Huang   University of Science and Technology Beijing, China
Hamidah Ibrahim   Universiti Putra Malaysia, Malaysia
Ivan Ivanov   SUNY Empire State College, USA
Cheqing Jin   East China Normal University, China
Konstantinos Kalpakis   University of Maryland Baltimore County, USA
Nikos Karacapilidis   University of Patras and CTI, Greece
Dimitris Karagiannis   University of Vienna, Austria
Maurice van Keulen   University of Twente, The Netherlands
Foutse Khomh   École Polytechnique, Canada
Benjamin Klöpper   ABB Corporate Research, Germany
Mieczyslaw Kokar   Northeastern University, USA
John Krogstie   NTNU, Norway
Martin Krulis   Charles University, Czech Republic
Raimondas Lencevicius   Nuance Communications, USA
Ziyu Lin   Xiamen University, China
Ricardo J. Machado   Universidade do Minho, Portugal
Zaki Malik   Wayne State University, USA
Tiziana Margaria   University of Limerick and Lero, Ireland
Florent Masseglia   Inria, France
Fabio Mercorio   University of Milano-Bicocca, Italy
Dimitris Mitrakos   Aristotle University of Thessaloniki, Greece
Stefano Montanelli   Università degli Studi di Milano, Italy
Mena Badieh Habib   University of Twente, The Netherlands
  Morgan
Erich Neuhold   University of Vienna, Austria
Boris Novikov   Saint Petersburg University, Russian Federation
José R. Paramá   Universidade da Coruña, Spain

Jianlong Zhong   GRAPHSQL INC, USA
Hong Zhu   Oxford Brookes University, UK
Yangyong Zhu   Fudan University, China

Additional Reviewers

Nikos Bikakis   National Technical University of Athens, Greece
Estrela Ferreira Cruz   Instituto Politécnico Viana do Castelo, Portugal
Matteo Francia   University of Bologna, Italy
Enrico Gallinucci   University of Bologna, Italy
Giorgos Giannopoulos   Institute for the Management of Information Systems,
                          Greece
Janis Kampars   Riga Technical University, Latvia
Ilias Kanellos   IMIS/Athena RC, Greece
Inese Polaka   Riga Technical University, Latvia
Dimitris Skoutas   RC Athena, Greece
Nikolaos Tantouris   University of Vienna, Austria
Konstantinos Zagganas   Institute for the Management of Information Systems,
                          Greece

Invited Speakers

Michele Sebag   Laboratoire de Recherche en Informatique, CNRS, France
John Domingue   The Open University, UK
Paul Longley   University College London, UK

Contents

Decision Support System for Implementing Data Quality Projects

Meryam Belhiah[✉], Mohammed Salim Benqatla,
and Bouchaïb Bounabat

Laboratoire AL QualSADI, ENSIAS, Mohammed V University in Rabat,
Rabat, Morocco
{meryam.belhiah,salim.benqatla}@um5s.net.ma,
bounabat@ensias.ma

Abstract. The new data-oriented shape of organizations inevitably imposes the need for the improvement of their data quality (DQ). In fact, growing data quality initiatives are offering increased monetary and non-monetary benefits for organizations. These benefits include increased customer satisfaction, reduced operating costs and increased revenues. However, regardless of the numerous initiatives, there is still no globally accepted approach for evaluating data quality projects in order to build the optimal business cases taking into account the benefits and the costs. This paper presents a model to clearly identify the opportunities for increased monetary and non-monetary benefits from improved data quality within an Enterprise Architecture context. The aim of this paper is to measure, in a quantitative manner, how key business processes help to execute an organization's strategy and then to qualify the benefits as well as the complexity of improving data, that are consumed and produced by these processes. These findings will allow to select data quality improvement projects, based on the latter's benefits to the organization and their costs of implementation. To facilitate the understanding of this approach, a Java EE Web application is developed and presented here.

Keywords: Cost/benefit analysis · Data accuracy · Data quality projects assessment · Business processes

1 Introduction

As business processes have become increasingly automated, nothing is more likely to limit and penalize the business processes' performance and overall quality than ignored data quality. What impacts daily operations, financial and business objectives, downstream analysis for effective decision making and end-user satisfaction [1], whether it is a customer, a citizen, an institutional partner or a regulatory authority. The problem of identification and classification of costs inflicted by poor data quality, has been given great attention in literature [2, 3], as well as the definition of approaches to measure Return On Investment (ROI) of data quality initiatives in both research [4] and industrial areas [5].

Even though the work cited above establishes the overall methodology for measuring the business value of data quality initiatives, it lacks generic and concrete

M. Helfert et al. (Eds.): DATA 2015, CCIS 584, pp. 1–16, 2016.
DOI: 10.1007/978-3-319-30162-4_1

metrics, based on cost/benefit analysis, that can be used by different organizations in order to facilitate the identification of opportunities for increased benefits before launching further analysis using additional KPI that are specific to each organization. The overall goal is not to improve data quality by any means, but to carefully plan data quality projects that are cost-effective and that will have the most positive impact. This guidance is particularly crucial for organizations with no or only little experience in data quality projects.

While it is difficult to develop a generic calculation framework to evaluate costs and benefits of data quality projects in money terms, the purpose of this paper is to find a suitable model to assess the positive impact of the improvement of quality of a data object used by a key business process alongside the implementation complexity. This is relevant because the positive impact and implantation complexity could be transformed to quantitative measures of monetary benefits and costs. The application of the proposed model is demonstrated on selected key business processes and business objects that are owned by the Department of Lands of Morocco and the results are presented here. The calculations are performed automatically via our Web application, which is an implementation of the proposed model.

The organization of this paper is addressed as follows: Sect. 2 presents data quality dimensions as well as data quality projects lifecycle, Sect. 3 describes the main steps of our model and presents the results from our case study; in Sect. 4, the discussion and future work are summarized.

2 Data Quality: Definition, Assessment and Improvement

2.1 Data Quality: Definitions and Assessment

Data quality may be defined as "the degree to which information consistently meets the requirements and expectations of all knowledge workers who require it to perform their processes" [6], which can be summarized by the expression "fitness for use" [1]. The term data quality dimension is widely used to describe the measurement of the quality of data. Even if the key DQ dimensions are not universally agreed amongst academic community, we can refer to Pipino et al. [7] who have identified 15 dimensions:

- Intrinsic: accuracy, believability, reputation and objectivity;
- Contextual: value-added, relevance, completeness, timeliness and appropriate amount;
- Representational and accessibility: understandability, interpretability, concise representation, accessibility, ease of operations and security.

All case studies that aimed at assessing and improving data quality have chosen a subset of data quality dimensions, depending on the objectives of the study [8–11]. Measurable metrics were then defined to score each dimension.

While it is difficult to agree on the dimensions that will determine the data quality, it is however possible, when taking users' perspective into account [12], to define a basic subset of key dimensions, including: accuracy, completeness and timeliness.

Accuracy. Accuracy is defined as "the closeness of results of observations to the true values or values accepted as being true" [7]. Wang et al. [1] define accuracy as "the extent to which data are correct, reliable and certified". The associated metric is as follows:

$$\frac{number\ of\ accurate\ values}{Total\ number\ of\ all\ values} \tag{1}$$

Completeness. Completeness specifies how "data is not missing and is sufficient to the task at hand" [13]. As completeness has often to deal with the meaning of null values, it may be expressed in terms of the "ratio between the number of non-null values in a source and the size of the universal relation" [11]. Completeness is usually associated to the metric below [14]:

$$\frac{number\ of\ non-null\ values}{Total\ number\ of\ all\ values} \tag{2}$$

Depending on the context, both accuracy and completeness may be calculated for: a relation attribute, a relation [15], a database or a data warehouse [16].

Timeliness. Timeliness is a time-related dimension. It expresses "how current data are for the task at hand" [13].

As a matter of fact, even if a data is accurate and complete, it may be useless if not up-to-date.

$$\frac{number\ of\ values\ that\ are\ up-to-date}{Total\ number\ of\ all\ values} \tag{3}$$

2.2   Data Quality Projects Lifecycle

Research in this area has shown that poor data quality is costing businesses a significant portion of their revenues. In the US, The Postal Service estimated it cost $1.5 billion in fiscal year 2013 to process undeliverable as addressed mail [17]. A 2011 report by Gartner [5], for instance, noted that as much as 40 % of the anticipated value of all business initiatives is never achieved due to poor data quality. In fact, poor data quality affects daily operations, labor productivity, management decision making and downstream analysis.

As such, companies have to evaluate different scenarios related to data quality projects to implement. The optimal scenario should provide the greatest business value and meet requirements regarding the available time, resources and cost.

Prior to introducing our model in the next section, it is suitable to present the common phases that compose a basic data quality project in an Enterprise Architecture context:

Define. Define the various dimensions of data quality from the perspective of the people using the data, using appropriate tools: survey studies, questionnaires, interviews, etc.

Measure. Associate data quality metrics to score each dimension.

Analyze. Interpret measurement results.

Improve. Design and implement improvement solutions on data and processes to meet requirements regarding the quality of data.

In an Enterprise Architecture context, it is also mandatory to perform process modeling. In fact, a piece of data represented by a business object is accessed in reading or writing modes by a process. Business objects produced by a process may serve as entry data to other downstream processes.

In a further step, it is also possible to perform data augmentation which consists of combining internal data with data from third parties to increase data coverage.

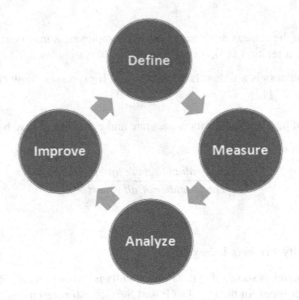

Fig. 1. Phases that compose a data quality project.

As cited above, we are particularly interested in assessing data quality projects that are related to specific dimensions of data quality, including: accuracy, completeness and timeliness. For the need of our case study, the remainder of this paper will focus on the accuracy dimension.

3 Model for Assessing Data Quality Projects

3.1 Business Processes' Positive Impact Assessment

The first part of our approach to track ROI of data quality projects consists of understanding how an organization's business/financial objectives and results are linked to key business processes' performance and overall quality. The following steps summarize the process of measuring the positive impact of the performance and overall quality of business processes on the strategy execution of an organization:

1. Identify leading factors that contribute to achieving short-term business/financial objectives of an organization;
2. Configure the importance of these factors according to the specifications of each organization;
3. Measure the impact of key business processes' performance and overall quality on these factors;
4. Order business processes by positive impact.

Leading Factors That Help Achieving Business/Financial Objectives of an Organization. To understand how business processes' performance and overall quality affect the success of an organization, financial/business objectives and results are detailed as follows:

- Positive impact on daily operations;
- Increasing revenues;
- Increasing productivity;
- Reducing costs;
- Meeting regulatory driven compliance;
- Positive impact on effective decision making;
- Positive impact on downstream analysis.

Configuration of Importance of the Above Factors According to the Specifications of Each Organization. Due to organizations' specific aspects and sets of success factors and in order to provide a generic approach that can be implemented without any adjustment, the second step of our approach introduces the context-aware and configurable weighting coefficients, illustrated in Table 1.

The purpose behind using a weighing coefficient is to allow each organization to express the importance of a success factor, depending on its context and strategy. To cite few examples where using different weighting coefficients is relevant:

- Public organizations may have more concerns about increasing end-users satisfaction (citizens in this particular case), than increasing revenues;
- Healthcare actors may give more attention to meeting regulatory driven compliance than to the other factors, while still important, owing to the fact that norms and standards are mandatory in the field of healthcare;
- Industrial companies may give the same importance to all the factors above.

Table 1. Configuration canvas for positive impact calculation.

Factor	Values	Rating (R)	Weighting coefficient (I)
Impact on daily operations	• True • False	1 0	
Impact on short-term business/financial objectives	• Increasing revenues • Increasing productivity • Reducing costs • Increasing end-user satisfaction • Meeting regulatory driven compliance • Other	0.15 0.15 0.15 0.15 0.15 0.15	
Impact on decision making	• True • False	1 0	
Is the process cross-functional?	• True • False	1 0	

Measurement of the Impact of Key Business Processes' Performance on Overall Quality. Business and IT leaders in charge of data quality initiatives should:

1. List all the key business processes;
2. Configure the importance of each factor by acting on the associated weighting coefficient. The sum of all weighing coefficient must be equal to 100;
3. For each factor in column 1, choose the corresponding value in column 2 and rating in column 3.

In the case of an organization with many key business processes, the positive impact of each business process will be calculated, using the weighted sum strategy:

$$\sum_{i=1}^{m}(Ri * Ii)/100 \tag{4}$$

Where R_i is the rating for the factor "i" and I_i is the weighing coefficient that is associated with the factor "i", that was previously defined by both business and IT leaders. The obtained score ranges between 0 and 5, where "0" refers to "unnoticed impact" and "5" refers to "high positive impact". The Table 2 depicts the correspondence between the positive impact score and the impact level.

Table 2. Positive impact levels.

Impact score	Impact level
0–1.5	0 – unnoticed to low impact
1.5–3	1 – medium impact
3–4.25	2 – high impact
>4.25	3 – very high impact

Order Business Processes by Positive Impact. After iterating over all key business processes and calculating the associated positive impact score, business processes are automatically classified by priority, in order to spot the point of departure to identify opportunities for increased benefits from improved data quality.

As business processes consume and produce data, classifying key business processes by positive impact on an organization's short-term objectives and results, should be followed by the identification of data quality options with the greatest business value at least-cost.

In addition to the positive impact score, other leading indicators may be assessed using the same approach, including: agile transformation of business processes and potential risks that are associated with data quality initiatives. These aspects will be explored in a future work.

Because business processes access data objects in reading and/or writing modes, it is normal that the quality of the data has an impact on the result of business processes' execution and vice-versa.

3.2 Implementation Complexity Assessment

While the first part of our approach deals with understanding and assessing how business processes' performance and overall quality positively impact an organization's objectives and results, the second part of our approach focuses on data that are consumed and used by these processes.

The following steps detail the process of scoring the implementation complexity of data accuracy improvement, what will allow to associate different levels of complexity (ranking from low to very high complexity) to DQ improvement projects:

1. Identify leading factors that contribute to the calculation of the implementation complexity of data accuracy improvement;
2. Configure the importance of these factors according to the specifications of each organization;
3. Measure the positive impact and the implementation complexity;
4. Prioritize data to improve according to the scores obtained in the previous step.

Leading Factors that Help Calculating the Implementation Complexity of Data Accuracy Improvement. Data profiling activities should allow answering the questions in Table 3.

In this part, the weighting coefficient plays the same role as in the previous part, as it allows taking into consideration the particularities of each organization.

Measurement of the Complexity of Data Accuracy Improvement Projects. For a given data used by a key business process, the implementation complexity will be calculated as follows:

Table 3. Configuration canvas for complexity calculation.

Factor	Values	Rating (R)	Weighting coefficient (C)
Are there standards to restructure and validate the data?	• **False** • **True**	1 0	
Is there an authentic source of data (repository) that allows to complement or contradict the data?	• **False** • **True**	1 0	
Does the data object have attributes with great weight identification in relation to another data source?	• **False** • **True**	0 1	
Is the data processing:	• **Manual** • **Semi-automatic** • **Automatic**	1 0.5 0.25	
What is the size of the data to process?	• **Very large** • **Large** • **Medium** • **Low**	1 0.75 0.5 0.25	

$$\sum_{i=1}^{m}(Ri * Ci)/100 \qquad (5)$$

Where R_i is the rating for the factor "i" and C_i is the weighing coefficient that is associated with the factor "i", that was defined previously by both business and IT leaders. The obtained score ranges between 0 and 5, here where "0" refers to "minimal complexity" and "5" refers to "severe complexity". The Table 4 shows the correspondence between the complexity score and the complexity level.

Table 4. Complexity levels.

Complexity score	Complexity level
0–1.5	0 – very low-to-low complexity
1.5–3	1 – medium complexity
3–4.25	2 – high complexity
>4.25	3 – very high complexity

The Fig. 2 presented below summarizes the main steps of our approach:

1. Determine what processes contribute the most to the business's objectives and results;
2. Determine data improvement complexity;
3. Recommend the optimal business case for data improvement.

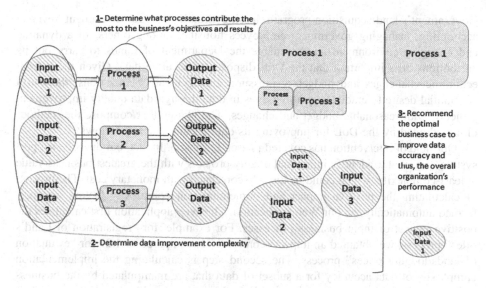

Fig. 2. Main phases.

After completing our research and in order to calculate automatically our indicators, we have implemented a Web application, which main functionalities are:

1. Create a new business process;
2. List registered business processes;
3. Add a new business object (physically implemented by a data object), that is used by a registered business process;
4. List registered business objects;
5. Assess data accuracy improvement projects;
6. List all previous assessments.

3.3 A Use Case Study: Decision Support System for Implementing DQ Projects in Action

For the purpose of verifying and validating the decision support system for implementing DQ projects, we have performed an analysis of data for Morocco Department of Lands (DoL)[1].

The Department of Lands administers the State-owned land mass. Its primary purpose is to unlock the potential of State's lands for the economic, social and environmental benefits, while optimizing the value of the State's land assets. The Department of Lands is pursuing its objectives through a number of fully automated business processes including: administering the sale of State's property, managing the

---

[1] http://www.domaines.gov.ma/.

Government's land acquisition program, leasing eligible lands for investment, revenue accounting, managing government employees housing, among others. In a dynamic and changing environment and to enable the Department of Lands to carry out its attributions, it is important that the DoL disposes of accurate data. Given the current economic challenges and budgetary pressures facing most organizations, there is a substantial desire to eradicate quality issues in data through data quality improvement projects, with reasonable budget and changes. The phases that compose the methodology adopted by the DoL for improving its data accuracy are as follows:

Our field of intervention has covered phases 1, 2 and 5. In fact, our decision support system was used to identify individual quality projects with the greatest business value at least-complexity, which could be directly correlated with monetary cost. The process of calculating the positive impact and implementation complexity factors was performed automatically via our Web application. The Web application first calculates the positive impact of input business processes. For example, for "registration of Land's titles" process, we obtained an impact score of 4.71, compared to 4.29 for "evaluation of lands/lodging prices" process. The second step is calculating the implementation complexity of data accuracy for a subset of data that are manipulated by the business processes. The third step is recommending the scenario with the greatest positive impact at least complexity and cost. The list of the selected business processes and data objects that were considered for our experimental setup where suggested by the users of DoL Information System.

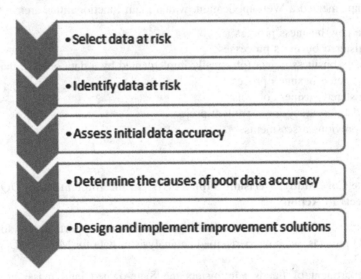

Fig. 3. Phases of the data quality improvement methodology adopted by the DoL.

Table 5 above describes the potential business objects that are considered for data accuracy improvement. Meanwhile, Table 6 below shows how these business objects are manipulated by critical business processes in terms of database operations. For each business object, the identifier is preceded by the symbol "#".

Table 5. Description of business objects.

Business object	Definition	Attributes
Evaluated price	• The result of the evaluation of lands that are owned by the State, in order to sell them or lease them for investment	• #Property_id • Evaluated price
Property –legal information	• Information that describe the legal situation of a land that is owned by the State, in terms of its type (registered, requisition, unregistered), owners and quota shares	• #Property_id • Legal situation • Owners • Quota shares
Property – urban situation	• Information that describe the urban situation of a land that is owned by the State, in terms of its area and geographical data	• #Property_id • Area • Zoning • Geographical coordinates
Procedure file	• Information that describe the procedure file, including its identifier, location, file opened and closed dates, as well as the related business procedure	• #File_id • Location • File opened date • File closed date • Business procedure
Accounting document	• Information that describe the type of the accounting document (revenue or expanse), its identifier, amount and related business procedure	• #Accounting_ document_id • Accounting_ document_type • Amount • Business procedure
Beneficiary	• Information that identify the natural or the legal person	• #Identifiers • Category of the beneficiary • Address

Choosing a Relevant Subset. Keeping in mind that this issue is a statistical challenge because of the large size of databases and/or data files, we have chosen a statistical method for determining the reliable sample size with given restrictions such as the margin of error and the confidence level. The sample size is represented by the Eq. 6 [18]:

$$n = \frac{Z^2 \times s \times (s-1)}{e^2} \qquad (6)$$

In Eq. 6, (n) is the sample size, (e) is the margin of error and (s) is the percentage of compliance between records in DoL internal databases and other authentic data sources. The margin of error indicates the accuracy of the chosen sample and the allowed deviation of the expected results. In our calculation, we used a 5 % value for the margin error. The confidence level tells how often the true percentage of the sampled data

Table 6. Access matrix.

Business object	Business process				
	Evaluation of lands/lodging	Registration of Lands titles	Litigation proceedings	Revenue accounting	Government Employee Housing
Evaluated price	CRUD*	-	read	read	read
Property – legal information	read	CRUD	read	-	read
Property – urban situation	read	CRUD	read	-	read
Procedure file	read	read	CRUD	read	read
Accounting document	-	-	-	CRUD	read
Beneficiary	read	-	read	read	CRUD

(*) refers to database operations: Create, Read, Update and Delete.

satisfying the required condition lies within the confidence interval. Usually, (Z) is chosen to be 90 or 95 %. For the latter value, Z takes a critical value of 1.96.

Decision Support Tool in Action. Our Web application was used to automatically calculate the positive impact and complexity indicators. The results of the experiment are provided below:

Figure 4 illustrates the results of the positive impact score calculation. As can be observed, the highest score was related to "registration of Lands titles" process followed by "revenue accounting". "Government Employee Housing" had the least impact. Figure 5 illustrates the results of the implementation complexity score calculation. As can be observed, the highest score was related to "procedure file" data object.

A closer look at the results of the questionnaires administered via our Web application provides information regarding the elements that contributed to high or low levels of positive impact and implementation complexity. Data accuracy improvement for the "procedure file" is very complex to setup owing to the fact that there are neither

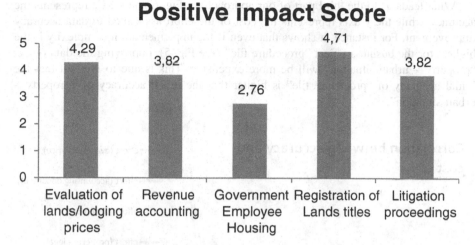

Fig. 4. Positive assessment score for the analyzed business processes.

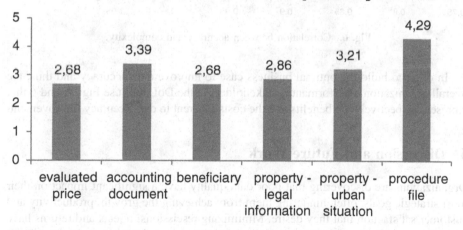

Fig. 5. Implementation complexity score for the analyzed business objects.

standards to restructure and validate the data nor an authentic source of data that allows to complement or contradict the data. Also the file processing is manual and the size of data to process is high.

What we have accomplished so far allows us to associate quantitative measures to the positive impact and the implementation complexity of data accuracy improvement projects. Meanwhile, when we try to estimate the costs that are associated with these projects, we should also take into consideration the initial accuracy of the business objects. This is relevant because it is less expensive to improve a data object with higher initial data accuracy.

What leads us to the third part of our approach: in Fig. 6, the x axis represents the accuracy while the y axis represents the cost or the effort associated to data accuracy improvement. For instance, it shows that even if the implementation complexity is the highest for the business object "procedure file" (see Fig. 5), improving the data object "property – urban situation" will be more expensive. This is due to the fact that the initial accuracy of "procedure file" is higher that the initial accuracy of "property – urban situation".

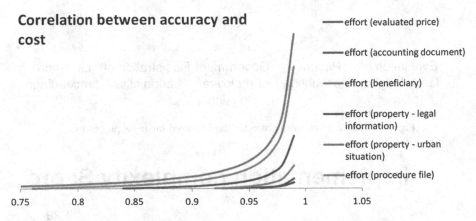

Fig. 6. Correlation between accuracy and complexity.

In order to build the optimal business case to improve data accuracy and thus, the overall organization's performance, stakeholders at the DoL will use Figs. 4 and 6 that represent respectively the benefits and the costs inherent to data accuracy improvement.

4 Discussion and Future Work

Organizations are discovering that poor data quality have a significant impact on their most strategic goals, often hindering them from achieving the growth, productivity and customer satisfaction that they desire. Minimizing rescissions, rejects and returns have positive impacts on daily operations, cost reduction and financial results. Thus, high-quality data are the precondition for leveraging processes run either by corporations or government agencies, in order to achieve their business initiatives. The demand for data quality assessment and improvement methodologies is maturing, especially when it comes to organizations having no or very little experience in the field of data quality projects.

Since the automation of business processes guarantees, in a way, the quality of their execution, actions must be directed towards the improvement of the accuracy of the data used by these processes. Our approach highlights the most cost-effective data accuracy improvement projects. We have established two global indicators of positive impact and implementation complexity, to measure the business value of data accuracy improvement projects. Furthermore and in order to recommend the optimal business case to improve data accuracy and thus, the overall organization's performance, our model takes into account: (1) – the initial data accuracy level (as-is), (2) – the positive impact of the key process that uses the data and (3) – the implementation complexity of data accuracy improvement initiative. According to the values of these indicators and to the targeted accuracy level (to-be), two business cases may be considered:

The first one is based on the improvement of data accuracy by determining and analyzing the sources of low quality, such as uncontrolled data acquisition, updates problems, etc.

The second one is process-driven as it encourages the improvement of processes (reengineering, control, etc.), by enhancing their execution accuracy. This is a short term option that is generally less expensive, but requires change management because it affects the work processes; In fact, while technology plays a key role in data quality improvement, changes in working methods are critical.

To summarize, the result of the work accomplished so far shows how to measure in a quantitative manner the business value of data quality improvement projects by establishing two global indicators of positive impact and implementation complexity.

As each organization environment is different, it is challenging to see how our model will perform in other contexts other than Enterprise Architecture environment. We are particularly interested in applying it to the context of Open Data that are produced by national governments, where data quality issue could derail the Open Data projects from their purpose.

References

1. Wang, R.Y., Strong, D.M.: Beyond accuracy: what data quality means to data consumers. J. Manage. Inf. Syst. **12**(4), 5–33 (1996)
2. Eppler, M., Helfert, M.: A classification and analysis of data quality costs. In: International Conference on Information Quality, pp. 311–325 (2004)
3. Haug, A., Zachariassen, F., Van Liempd, D.: The costs of poor data quality. J. Ind. Eng. Manage. **4**(2), 168–193 (2011)
4. Otto, B., Hüner, K. M., Österle, H.: Identification of business oriented data quality metrics. In: ICIQ (2009)
5. Gartner.: measuring the business value of data quality (2011). https://www.data.com/export/sites/data/common/assets/pdf/DS_Gartner.pdf
6. International Association for Information and Data Quality (2015). http://iaidq.org/main/glossary.shtml
7. Pipino, L.L., Lee, Y.W., Wang, R.Y.: Data quality assessment. Commun. ACM **45**(4), 211–218 (2002)
8. Aladwani, A.M., Palvia, P.C.: Developing and validating an instrument for measuring user-perceived web quality. Inf. Manage. **39**(6), 467–476 (2002)
9. Batini, C., Comerio, M., Viscusi, G.: Managing quality of large set of conceptual schemas in public administration: methods and experiences. In: Abelló, A., Bellatreche, L., Benatallah, B. (eds.) MEDI 2012. LNCS, vol. 7602, pp. 31–42. Springer, Heidelberg (2012)
10. Scannapieco, M., Catarci, T.: Data quality under a computer science perspective. Arch. Comput. **2**, 1–15 (2002)

11. Närman, P., Johnson, P., Ekstedt, M., Chenine, M., König, J.: Enterprise architecture analysis for data accuracy assessments. In: Enterprise Distributed Object Computing Conference (2009)
12. Belhiah, M., Bounabat, B., Achchab, S.: The impact of data accuracy on user-perceived business service's quality. In: 10th Iberian IEEE Conference on Information Systems and Technologies (2015)
13. Batini, C., Scannapieco, M.: Data Quality: Concepts, Methodologies and Techniques. Springer, Heidelberg (2006)
14. Bovee, M., Srivastava, R.P., Mak, B.: A conceptual framework and belief function approach to assessing overall information quality. Int. J. Intell. Syst. 18(1), 51–74 (2003)
15. Naumann, F.: Quality-Driven Query Answering for Integrated Information Systems. LNCS, vol. 2261. Springer, Heidelberg (2002)
16. English, L.P.: Improving Data Warehouse and Business Information Quality: Methods for Reducing Costs and Increasing Profits. Wiley, New York (1999)
17. Office of Inspector General/United States Postal Office. Audit report: Undeliverable as Addressed Mail. MS-AR-14-006 (2014)
18. NIST/SEMATECH. E-Handbook of statistical methods (2013). http://www.itl.nist.gov/div898/handbook/

Classify Visitor Behaviours in a Cultural Heritage Exhibition

Salvatore Cuomo[1]([✉]), Pasquale De Michele[1], Ardelio Galletti[2],
and Giovanni Ponti[3]

[1] Department of Mathematics and Applications,
University of Naples "Federico II", Naples, Italy
{salvatore.cuomo,pasquale.demichele}@unina.it
[2] Department of Science and Technology,
University of Naples "Parthenope", Naples, Italy
ardelio.galletti@uniparthenope.it
[3] Technical Unit for ICT – High Performance Computing Division,
ENEA – Portici Research Center, Portici, Italy
giovanni.ponti@enea.it

Abstract. Classify the dynamic of users in a cultural heritage exhibition in order to infer information about the event fruition is a very interesting research field. In this paper, starting from real data, we investigate the user dynamics related to the interaction with artworks and how a spectator interacts with available technologies. Accordingly with the fact that the technology plays a crucial role in supporting spectators and enhancing their experiences, the starting point of this research has been the art exhibition named *The Beauty or the Truth* that was located in Naples (Italy), where event was equipped with several technological tools. Here, the collected log files, stored in a suitable expert software system, are used in a flexible framework in order to analyse how the supporting pervasive technology influence and modify behaviours and visiting styles. Finally, we carried out some experiments to exploit the clustering facilities for finding groups that reflect visiting styles. The obtained results have revealed interesting issues also to understand hidden aspects in the data and unattended in the analysis.

Keywords: User profiling · Clustering · Data mining

1 Introduction

In the cultural heritage area, the requirements of innovative tools and methodologies to enhance the quality of services and to develop smart applications is an increasing requirement. Cultural heritage systems contain a huge amount of interrelated data that are more complex to classify and analyse. For example, considering an art exhibition, characterizing, studying, and measuring the level of knowledge of a visitor with respect to an artwork, and also the dynamics

M. Helfert et al. (Eds.): DATA 2015, CCIS 584, pp. 17–28, 2016.
DOI: 10.1007/978-3-319-30162-4_2

of social interaction on a relationship network is an interesting research scenario. To understand and analyse how artworks observation can influence the social behaviours is a very hard challenges. Indeed, semantic web approaches have been increasingly used to organize different art collections not only to infer information about a cultural item, but also to browse, visualize, and recommend objects across heterogeneous collections [8]. Other methods are based on statistical analysis of user datasets in order to identify common paths (i.e., *patterns*) in the available information. Here, the main difficulty is the management and retrieval of large databases as well as issues of privacy and professional ethics [7]. Finally, models of artificial neural networks, typical of Artificial Intelligence field are also adopted. Unfortunately, these approaches seems to be, in general, too restrictive in describing complex dynamics of social behaviours and interactions in the Cultural Heritage framework [6].

In our previous works, we referred to a computational neuroscience terminology for which a cultural asset visitor is *a neuron* and its interest is *the electrical activity* which has been stimulated by appropriate currents. In detail, we adopted two different strategies to perform data classification: a Bayesian classifier [3], and an approach that finds data groupings in an unsupervised way [2]. Such a strategy resorts to a *clustering* task employing the well-known K-means algorithm [5]. The dynamics of the information flows, which are the social knowledge, are characterized by neural interactions in biological inspired neural networks. Reasoning by similarity, the users can be considered as neurons in a network and their interests the morphology; the common topics among users are the neuronal synapses; the social knowledge is the electrical activity in terms of quantitative and qualitative neuronal responses (spikes).

In [1] we dealt with the characterization of visitor behaviours starting from real datasets. As a real scenario, we have considered the art exhibition named *the Beauty or the Truth* located in Naples, Italy, where new ICT tools and methodologies, producing several users behavioural data, have been deployed and currently are still active. Our aim was also be to classify visiting styles of the spectators in the exhibit exploiting their user experience of the adopted technology, which allows to collect the data to analyse. Here, deeply investigating these aspects, we focus on tuning of the parameters influencing the classification results.

The paper is organized as follows. In Sect. 1 we introduce this research; in Sect. 2 we describe the fruition system framework of the proposed event. In Sect. 3 we deal with the visiting styles by resorting some basics definitions; in Sect. 4 we describe experiments on real data classification and finally in Sect. 5 we draw the conclusions.

2   The Digital Fruition System

The starting point of our research is an interesting and wide case study; it consists of a real art exhibition of 271 sculptures, divided into 7 thematic sections and named *"The Beauty or the Truth"*[1]. This exhibition shows, for the first time

---

[1] http://www.ilbellooilvero.it.

in Italy, the Neapolitan sculpture of the late nineteenth century and early twentieth century, through the major sculptors of the time. The sculptures are exhibited in the beautiful monumental complex of *San Domenico Maggiore*, in the historical centre of Naples. In order to better understand motivations behind this work, it is important to deeply analyse the kind of relations that exist among cultural spaces, people and technological tools that nowadays are pervasive in such environments. Accordingly, the behaviour of a person/visitor, when it is immersed inside a space and consequently among several objects, has to be analysed in order to design the most appropriate ICT architecture and to establish the relationship between people and technological tools that have to be non-invasive. For this reason, it should be preferable to provide cultural objects with the capability to interact with people, environments and other objects, and to transmit the related knowledge to users through multimedia facilities. In an intelligent cultural space, technologies must be able to connect the physical world with the world of information, in order to amplify the knowledge, but also and especially the fruition, involving the visitors as active players to which is offered the pleasure of the perception and the charm of the discovery of a new knowledge.

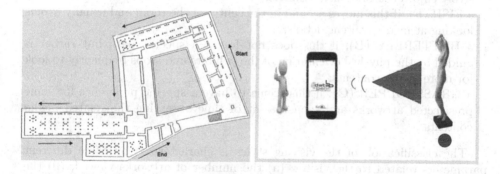

Fig. 1. Digital fruition system.

In the follow, the architecture of an Internet of Things (IoT) system, the technological sensors immersed in the cultural environment and the communication framework are presented. The sensors aimed to transform cultural items in *smart objects*, that now are able to communicate with each other, the visitors and the network; this acquired identity plays a crucial role for the smartness of a cultural space. Accordingly, in order that this system can perform its role and improve end-users cultural experience transferring knowledge and supporting them, a mobile application has been designed; in this way people have the opportunity to enjoy the cultural visit and be more at ease simply using their own mobile device. The proposed IoT system was entirely deployed inside the exhibition, as illustrated in Fig. 1. In the proposed system we have, on the left of Fig. 1 a physical space where the exhibition was located. It is important to observe that the combination of the technologies, artworks and path inside the event influences the visiting style of a spectator. On the right of Fig. 1 we have a detailed description of the interaction of the user with the artwork. The artworks start to talk and interact with the spectator through an "ad hoc" technology named smart cricket.

The overall technology in the proposed fruition system is used in order to collect the data that are necessary to characterize the visiting styles discussed in the following sections.

3   Visiting Styles Definition

In order to classify visiting styles in art exhibition, we start from work shown in [9], where personalized information presentation in the context of mobile museum guides are reported and visitor movements are classified comparing these to the behaviours of four typical animals. In our work, we adapt this classification to find how visitors interact with the ICT technology. Accordingly, the visitor's behaviour can be compared to that of:

- an ANT (**A**), if this tends to follow a specific path in the exhibit and intensively enjoys the furnished technology;
- a FISH (**F**), if this moves around in the centre of the room and usually avoids looking at media content details;
- a BUTTERFLY (**B**), if this does not follow a specific path but rather is guided by the physical orientation of the exhibits and stops frequently to look for more media contents;
- a GRASSHOPPER (**G**), if this seems to have a specific preference for some preselected artworks and spends a lot of time observing the related media contents.

The classification of the visiting styles is characterized by three different parameters related to the visitor: (a) the number of artworks viewed, (b) the average time spent by interacting with the viewed artworks, and (c) the path determining the order of visit of the exhibit sections.

Table 1. Characterization of the visiting styles' classification.

Animal	(a) Viewed artworks	(b) Average time	(c) Path
A	high	-	high
B	high	-	low
F	low	low	-
G	low	high	-

As we can observe in Table 1, high values for the parameter (a) characterize both **A**s and **B**s, while low values are related to **F**s and **G**s. Moreover, the parameter (b) does not influence the classification of **A**s and **B**s, while high values are typical for **F**s and low values are inherent in **G**s. Finally, the parameter (c) does not influence the classification of **F**s and **G**s, whereas high values are related to **A**s and low values characterize **B**s.

Each parameter is associated with a numerical value normalized between 0 and 1. For the parameters (a) and (b), its values respectively correspond to percentages of viewed artworks and average time spent for these. In the following we show how to assign a value to the parameter (c).

Assume that the N sections (rooms) of a museum are organized in increasing order as follows:

$$\text{entrance}\ \boxed{1} \to \boxed{2} \to \ \cdots\ \to \boxed{\text{N-1}} \to \boxed{\text{N}}\ \text{exit},$$

Let us denote by

$$S_1 \to S_2 \to \ \cdots\ \to S_{M-1} \to S_M$$

the path of a visitor. Assuming the entrance in 1 and the exit in N, it results $S_1 = 1$ and $S_M = N$. Moreover, because a visitor could visit each room more than once or never, the length M of the path is independent on the number N of rooms. To assign a value to the quality of a path, we consider the $M - 1$ movements $S_{j-1} \to S_j$, for $j = 2, \ldots, M$, and we assign at each of them the value m_j in the following way

$$m_j = \begin{cases} 1, & \text{if } S_j = S_{j-1} + 1 \\ 0.5, & \text{if } S_j > S_{j-1} + 1 \\ S_j - S_{j-1}, & \text{if } S_j < S_{j-1} \end{cases}$$

Finally, the value of the path will be

$$p = \max\left\{0, \frac{1}{N-1}\sum_{j=2}^{M} m_j\right\}.$$

The ratio by $N - 1$ ensures that $p \leq 1$, while the max operation avoids negative values, so that the property $0 \leq p \leq 1$ is achieved. For example, with $N = 7$ rooms, the *chaotic* path

$$\textit{entrance}\ \boxed{1}\ \underset{m_2=0.5}{\to}\ \boxed{3}\ \underset{m_3=1}{\to}\ \boxed{4}\ \underset{m_4=0.5}{\to}$$

$$\boxed{6}\ \underset{m_5=-4}{\to}\ \boxed{2}\ \underset{m_6=0.5}{\to}\ \boxed{6}\ \underset{m_7=1}{\to}\ \boxed{7}\ \textit{exit},$$

will be evaluated

$$p = \max\left\{0, \frac{1}{6}(0.5 + 1 + 0.5 - 4 + 0.5 + 1)\right\}$$

$$= \max\left\{0, \frac{-0.5}{6}\right\} = 0.$$

While the more ordered path

$$entrance \; \boxed{1} \; \underset{m_2=1}{\rightarrow} \; \boxed{2} \; \underset{m_3=1}{\rightarrow} \; \boxed{3} \; \underset{m_4=1}{\rightarrow} \; \boxed{4} \; \underset{m_5=0.5}{\rightarrow}$$

$$\boxed{6} \; \underset{m_6=1}{\rightarrow} \; \boxed{7} \; exit,$$

will be evaluated

$$p = \max\left\{0, \frac{1}{6}(1+1+1+0.5+1)\right\}$$

$$= \max\left\{0, \frac{4.5}{6}\right\} = 0.75$$

Observe that, in general, the perfect path

$$entrance \; \boxed{1} \rightarrow \boxed{2} \rightarrow \cdots \rightarrow \boxed{N\text{-}1} \rightarrow \boxed{N} \; exit,$$

will always receive the evaluation $p = 1$.

Observe that our model infers information from data stored in the structured JSON file. The overall data collected by the described ICT framework will be used as the input of the proposed classification approach. More in details, *LOG* files are structured in order to store main informations about the visitor behaviour in the exhibit. A listing that shows the JSON schema diagram of a log file, characterized by the fruition information w.r.t. the artworks is discussed in [1].

4 Experiments on Data Classification

The experiments described in this Section were carried out from a dataset of 253 log files, one for each visitor of the exhibition. Experiments have been executed exploiting the computing resources provided by CRESCO/ENEAGRID High Performance Computing infrastructure [4]. We have tracked the visitor behaviour by using a suitable Extrapolation Algorithm (EA), which has a JSON file as input data. A typical EA output is shown in the following:

```
IDUser :  e7a5774700c1e88e1417618582735
# of artworks:  271
# of viewed artworks:  44
% of viewed artworks :  17.5%


...
-------------------------------------
i-th viewed artwork :  2
ID artwork :  128
Available audio (sec.) :  32.922
Listen audio (sec.) :  32.922
Available images :  3
Viewed images :  0
```

```
Available text : True
Viewed text :  False
Interaction time (sec.) :   58.259
Path is followed :  True
------------------------------------------
...
------------------------------------------
i-th viewed artwork :  6
ID artwork :  17
Available audio (sec.) :  85.141
Listen audio (sec.) :  85.141
Available images :  4
Viewed images :  2
Available text : True
Viewed text :  True
Interaction time (sec.) :  103.141
Path is followed :  False
------------------------------------------
```

Such files are particularly suitable to identify user behaviour not only regarding their interactions with artworks, but also w.r.t. the whole artwork exhibition. In fact, properly looking at the JSON files, for each user it is possible to determine if the exhibition path is followed, the sequence of visited sections, the time spent to enjoy audio and image contents, and if text information about a specific artwork are visualized or not.

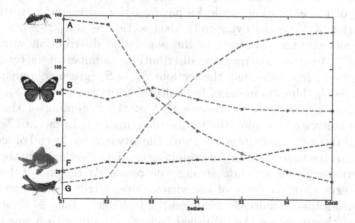

Fig. 2. Visitor classification starting from the section S_1 to all other sections (number of visitors).

Starting from the data collected in the exhibit, we deployed a classifier in order to characterize the visiting behaviours by means of some heuristics and to investigate how visitors interact with the supporting technology. According with

Table 2. Visitor classification starting from the section S_1 to all other sections (in %).

Animal	$S_1 \rightarrow S_1$	$S_1 \rightarrow S_2$	$S_1 \rightarrow S_3$	$S_1 \rightarrow S_4$	$S_1 \rightarrow S_5$	$S_1 \rightarrow S_6$	$S_1 \rightarrow S7$
A	137	133	79	52	35	21	14
B	21	28	27	30	31	38	43
F	83	76	85	76	69	68	68
G	12	16	62	95	118	126	128

this and with the characterization of the classification illustrated in Table 1, we defined thresholds to better identify the three parameters of our dataset. In a preliminary phase, we set 10 % for viewed artwork, 60 % for average time, and 70 % for path.

Based on these thresholds, Fig. 2 and Table 2 show, respectively, the distribution number and percentage of the visitor style in all the exhibit sections calculated by our classifier. It is easy to note that **A**s and **G**s have specular trends among the sections, since **A**s start with a high population in the first two sections (i.e., 53 % moving from S_1 to S_2) but drastically decrease up to the 6 % at the end of the exhibit, whereas **G**s are thinly populated at the beginning of the exhibit but there is a strong increment starting from section S_3 (i.e., 25 %) that culminates at the end of the exhibit (i.e., 51 %).

On the other hand, we observe that **B**s and **F**s population trends are quite similar since tend to remain stable for the entire duration of the exhibition. In fact, **B**s have a max increment equal to 8 % from beginning (i.e., 8 %) to end (i.e., 16 %) of the exhibition, while **F**s have a 6 % of decrement starting from section S_1 (i.e., 33 %) and moving to the last section S_7 (i.e., 27 %).

We characterize the variations in the population distribution as metamorphosis. In Fig. 3 we summarize these distributions (number of visitors) for the sections $S_1 \rightarrow S_2$ (red columns), the sections $S_1 \rightarrow S_4$ (green columns) and the sections $S_1 \rightarrow S_7$ (blue columns) of the cultural heritage event.

From these experiments we deduce that at the beginning of the exhibit, visitors are motivated to enjoy the furnished technology. In fact, 64 % of these (i.e., **A**s and **B**s) intensively interacted with the artworks at the end of section S_2. Afterwards, as the time spent in the exhibit grows (i.e., starting from section S_3), the interaction with the artworks strongly decreases. At the end of the exhibit, we can observe that the 78 % of the visitors are distributed in two different groups: visitors that definitively abandon the technology (i.e., 27 % of **F**s) and visitors that choose to use the furnished technology only with a small number of specific artworks (i.e., 51 % of **G**s).

We performed a further experimental phase based on clustering algorithm. The aim of this step consists in executing an unsupervised data mining algorithm in order to achieve data groups that can reflect the user classification obtained with our classifier. In this direction, we propose a data structure that reflects the above mentioned observations about path and media content fruition. The dataset is built from the log files and is structured in ARFF Weka format, as shown in the following.

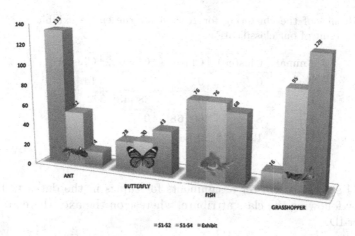

Fig. 3. Visitor metamorphosis for exhibit sections $S_1 \rightarrow S_2$, $S_1 \rightarrow S_4$ and the entire exhibit ($S_1 \rightarrow S_7$) (Color figure online).

```
@RELATION ARTWORKS
@ATTRIBUTE viewed NUMERIC [0..1]
@ATTRIBUTE avg_time NUMERIC [0..1]
@ATTRIBUTE path NUMERIC [0..1]
@ATTRIBUTE class {A,B,F,G}
@DATA
...
0.14741,1,1,A
0.0517928,0.94078,1,G
0.0119522,0.1273,1,F
0.135458,1,0.166667,B
...
```

In particular, the dataset contains fields regarding the percentage of viewed artworks (i.e., `viewed`), the average time spent in artwork interaction (i.e., `avg_time`), and the percentage of followed path during the visit (i.e., `path`). Note that these three attributes reflect the parameters introduced in Sect. 3, used to characterize the animal behaviours.

We have resorted to the well-known K-means partitional clustering algorithm [5] and set the number of classes to $K = 4$.

In Table 3 we report the results of the clustering (with $K = 4$) for the entire exhibit. Note that **Cluster0** corresponds to **G**, **Cluster1** is **F**, **Cluster2** represents **B** and **Cluster3** is **A**, as this is a typical majority voting based cluster assignment. This clustering session provides very interesting results that can be seen in the Table. In fact, the four categories in our data have been correctly identified by our classifier with an accuracy very close to the K-means results, with a number of incorrectly clustered instances equal to about 5 %.

Table 3. Results of the clustering for $K = 4$ for the entire exhibit $(S_1 \rightarrow S_7)$. In brackets the result of our classification.

Animals	Cluster0	Cluster1	Cluster2	Cluster3
A	0	0	0	14 [**14**]
B	0	0	38 [**43**]	5
F	8	60 [**68**]	0	0
G	127 [**128**]	0	0	1

Figure 4 shows the cluster assignments for tuples in the dataset. These are coloured by following the class attribute, whereas on the axes there are class-ID and cluster-ID.

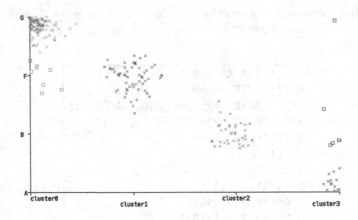

Fig. 4. Kmeans cluster assignment $(K = 4)$.

Clustering results confirmed what we addressed in the classification task, as the algorithm correctly identify classes **A** and **G**, achieving a very high accuracy results. Note that only one instance has a bad clustering assignment for **G**, whereas all the instances correctly clustered for **G**. The poor clustering error is due to the classes **B** and **F**, with the error of 5 tuples for **B** and 8 tuples for **F**. This is not a surprising result, since we yet noticed in the previous classification task that **B** and **F** are have very similar trends.

4.1 Tuning of the Parameters

In the previous experimental phase, we obtained very interesting results in terms of accuracy, with a very low clustering error (i.e., 5 %). However, such results came from an our intuition in setting proper thresholds in classifying visitor behaviours. In this section, we focus on the tuning of the classification parameters, in order to show how they bias the cluster accuracy.

We described in Sect. 3 how setting thresholds in order to classify visiting styles reflecting the ethnographic behaviour. Based on a possible range of values for each parameter, we defined a set of experiments in which the classification changes and the clustering strategy/setting remains the same. The goal here is to discover the best setting for the classification that produces the highest accuracy result in clustering phase for the entire exhibit ($S_1 \rightarrow S_7$).

We found several clustering accuracy results. In Table 4 we show only three most relevant setting, which produce the best, the medium and the worst quality results.

Table 4. Tuning of the classification parameters.

Quality	Parameters (%)			Accuracy error (%)	# Misclassified
	(a)	(b)	(c)		
Best	10 %	50 %	60 %	1.19	3
Medium	10 %	65 %	75 %	10.67	27
Worst	20 %	60 %	70 %	35.57	90

In Table 4, the first column expresses the quality evaluation of the tuning and the following three columns represent the settings for the three classification parameters, that are "viewed" artworks (a), "average time" (b), and "path" (c). We can notice that the first setting produces the best accuracy in clustering (1.19 % error). However, this result is very close to the one we obtained without the tuning (5 % error), and this confirms that our preliminary intuitions were right. In average, we had several parameter configurations which produce an error of about 10 %, which corresponds to a very good error rate in clustering. The worst setting produces an error of about 35 %, which is acceptable in clustering especially if we consider that our dataset is not so big (i.e., 253 tuples).

5   Conclusions

In this paper we have dealt with the user dynamics and behaviours starting from real datasets. Our aim has been to classify visitor visiting styles starting from data collected by the available technology. In order to validate the classification results, we resorted to the well-known K-means clustering algorithm to discover data groups reflecting visitor behaviours in all the sections of the exhibit. Experimental results have shown that clustering approach correctly identify visitor behaviours, providing high accuracy clusters that reflect the classification results. We have observed visitor behaviour modifications at the different exhibition sections, and this introduces the concept of metamorphosis in the visiting styles. An interesting observation and challenge for future works is to adapt, in a smart way, this computational framework to many different application topics, such as the context-aware profiling, feedback based and/or recommendation systems.

Acknowledgements. Authors thank DATABENC, a High Technology District for Cultural Heritage management of Regione Campania (Italy), and ENEA Portici Research Center, ICT-DTE-HPC Department and CRESCO HPC Cluster, for supporting the paper.

References

1. Cuomo, S., De Michele, P., Galletti, A., Ponti, G.: Visitor dynamics in a cultural heritage scenario. In: Proceedings - 4th International Conference on Data Management Technologies and Applications (DATA) (2015)
2. Cuomo, S., De Michele, P., Ponti, G., Posteraro, M.: A clustering-based approach for a finest biological model generation describing visitor behaviours in a cultural heritage scenario. In: Proceedings - 3rd International Conference on Data Management Technologies and Applications (DATA), pp. 427–433 (2014)
3. Cuomo, S., De Michele, P., Posteraro, M.: A biologically inspired model for describing the user behaviors in a cultural heritage environment. In: Proceedings - 22nd Italian Symposium on Advanced Database Systems (SEBD), pp. 292–302 (2014)
4. Ponti, G., et. al: The role of medium size facilities in the HPC ecosystem: the case of the new CRESCO4 cluster integrated in the ENEAGRID infrastructure. In: Proceedings of the International Conference on High Performance Computing and Simulation (HPCS), pp. 1030–1033 (2014)
5. Jain, A.K., Dubes, R.C.: Algorithms for Clustering Data. Prentice-Hall, Upper Saddle River (1988)
6. Kleinberg, J.: The convergence of social and technological networks. Commun. ACM **51**(11), 66–72 (2008)
7. Kumar, R., Novak, J., Tomkins, A.: Structure and evolution of online social network. In: Yu, P.S., Han, J., Faloutsos, C. (eds.) Link Mining: Models, Algorithms, and Applications, pp. 337–357. Springer, New York (2010). J. Am. Soc. Inf. Sci. Technol.
8. Middleton, S.E., Shadbolt, N.R., De Roure, D.C.: Capturing interest through inference and visualization: ontological user profiling in recommender systems. In: Proceedings - 2nd International Conference on Knowledge Capture, pp. 62–69 (2003)
9. Zancanaro, M., Kuflik, T., Boger, Z., Goren-Bar, D., Goldwasser, D.: Analyzing museum visitors' behavior patterns. In: Conati, C., McCoy, K., Paliouras, G. (eds.) UM 2007. LNCS (LNAI), vol. 4511, pp. 238–246. Springer, Heidelberg (2007)

An Automatic Construction of Concept Maps Based on Statistical Text Mining

Aliya Nugumanova[1](✉), Madina Mansurova[2],
Ermek Alimzhanov[2](✉), Dmitry Zyryanov[1], and Kurmash Apayev[1]

[1] D. Serikbayev East Kazakhstan State Technical University,
Ust-Kamenogorsk, Kazakhstan
yalisha@yandex.kz, {dzyryanov,kapaev}@ektu.kz
[2] Al-Farabi Kazakh National University, Almaty, Kazakhstan
mansurova01@mail.ru, aermek81@gmail.com

Abstract. In this paper, we explore the task of automatic construction of concept maps for various knowledge domains. We propose a simple 3-steps algorithm for extraction of key elements of a concept map (nodes and links) from a given collection of domain documents. Our algorithm manipulates a statistical term-document matrix describing how frequently terms occur in documents of the collection. At the first step we decompose this matrix into scores (terms-by-factors) and loadings (factors-by-documents) matrixes using non-negative matrix factorization, wherein each factor represents one topic of the collection. Since the scores matrix specifies the relative contribution of each term to the factors, we can select the most contributing terms and use them as concept map nodes. At the second step we associate selected key terms with the corresponding row-vectors of the term-document matrix and calculate pairwise cosine distances between them. Since the close distances determine the pairs of strongly related key terms, we can select the strongest relations as concept map links. Finally, we construct the resulting concept map as a graph with selected nodes and links. The benefits of our statistical algorithm are its simplicity, efficiency and applicability to any domain, any language and any document collection.

Keywords: Concept map · Text mining · Co-occurrence analysis · Non-negative matrix factorization

1 Introduction

Concept maps are graphical tools for representing knowledge structures of various domains. The main elements of concept maps are:

- Nodes (key concepts of the domain, put in circles or boxes);
- Links (key relations between concepts, represented as lines);
- Labels (words or phrases describing the meaning of relations).

The main purpose of concept maps is to contribute to a deeper understanding of domain knowledge on the conceptual level. The work [1] reports the results of experimental investigations which verify the practical value and efficiency of concept

M. Helfert et al. (Eds.): DATA 2015, CCIS 584, pp. 29–38, 2016.
DOI: 10.1007/978-3-319-30162-4_3

maps as a tool and as a strategy of teaching. Unfortunately, the complexity of a manual construction of concept maps greatly reduces the advantages of their using in the educational process. It is a very common case when teachers preparing course material are forced to use simple and limited types of concept maps or not use them at all because their comprehensive construction and drawing takes a lot of time.

Owing to the mentioned arguments, of great importance is the task of automatic or semi-automatic construction of concept maps on the basis of extraction their elements from collections of textual materials. Thanks to example of the authors of work [2] this task got the name Concept Map Mining (CMM) similar to Data Mining and Text Mining. In general case the process of CMM consists of three subtasks: extraction of concepts, extraction of links and summarization (see Fig. 1) [3].

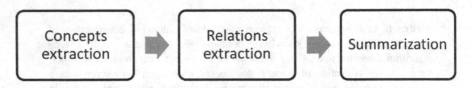

Fig. 1. The subtasks of concept map mining process.

The aim of this paper is to demonstrate usefulness and efficiency of statistical Text Mining methods for automatic construction of concept maps based on the domain collections of texts. The most important advantage of statistical methods is that they can be directly applied to any domain and any language, i.e. they are invariant in regard to the features of the given collection of domain documents. Statistical methods considered in this paper are based on the analysis of co-occurrence of terms in domain documents. We use a simple 3-steps algorithm which deals with a term-document co-occurrence matrix describing the number of occurrences of each word in each document of the collection.

At the first step we decompose the co-occurrence matrix into scores (terms-by-factors) and loadings (factors-by-documents) matrixes using non-negative matrix factorization, wherein each factor represents one topic of the collection. Since the scores matrix specifies the relative contribution of each term to the factors, we can select the terms with maximum contributions and use them as concept map nodes. At the second step we associate selected key terms (also known as concepts or nodes) with the corresponding row-vectors of the term-document matrix and calculate pairwise cosine distances between them. Cosine distance or similarity is a measure of similarity between two vectors that measures the cosine of the angle between them. We use this measure in positive space, where the outcome is bounded in [0,1]. So the maximum value of cosine similarity is equal to 1 (it corresponds to the angle 0). Since the close distances determine the pairs of strongly related key terms, we can select the strongest relations with similarity values more than 0.5 as concept map links. Finally, we construct the resulting concept map as a graph with selected nodes and links.

We plan to integrate automatically created concept maps into e-learning environment as a special tool supporting student's active and deep learning of the subject.

In [4] we represent the conception of our e-learning environment, and in this paper we investigate one of its meaningful elements.

The remaining part of the paper has the following structure. The second section presents a brief review of works related to the considered problem of automatic construction of concept maps. The approach proposed in the paper is described in detail in the third section. The results of experimental testing of the proposed approach are given in the fourth section. The fifth section contains brief conclusions on the work done and presents a plan of further investigations.

2 Related Works

The resent decade is characterized by the growth of interest to investigations devoted to automatic extraction of concept maps from collections of text materials. Among these studies, of high rank are the works based on the use of statistical techniques of processing a natural language. As is mentioned in [5], the methods focused on statistical processing of texts are simple, efficient and well portable; however, they possess a decreased accuracy as they do not consider latent semantics in the text.

The mentioned simplicity and efficiency of statistical approaches are illustrated well in [6]. The authors construct a term-term matrix based on a short list of key words selected manually for the given domain. They fill in the matrix on the basis of terms co-occurrences in sentences. If two elements occur in one sentence, the matrix element is equated to 1, otherwise – to 0. Then they display this matrix in the concept map, as shown in Fig. 2. Obviously, this approach is good for chamber teaching courses consisting of materials limited in volume, but for weighty courses it is very inefficient. The authors applied their methodology for constructing concept maps based on students' text summaries. Obtained concept maps were used by instructor to analyze how students learned the training material. In particular, the purpose of the analysis was selection of correct, incorrect or missing propositions in the students' summaries.

- I have two pets, my dog is named Buddy and my cat is named Missy.
- My dog likes to ride in my dad's truck.
- But not Missy (metonym for cat), she will only ride in my mom's car.

	Cat	Dog	Pet	Car	Truck
Cat	-				
Dog	1	-			
Pet	1	1	-		
Car	1	0	0	-	
Truck	0	1	0	0	-

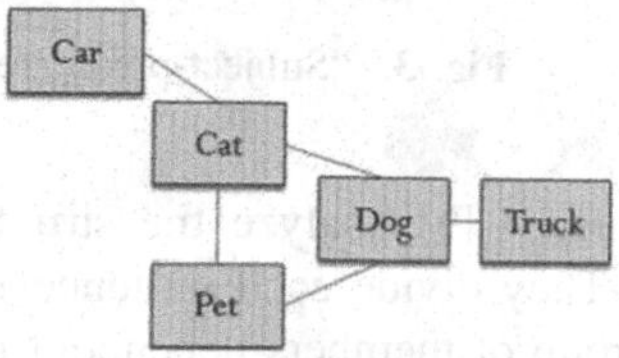

Fig. 2. Mapping the term-by-term matrix to the concept map (by work [6]).

The authors of [7] extract concepts from scientific articles using the principal component analysis. They use some papers in scientific journals and conference proceedings, dedicated to the field of e-learning, as data sources for the construction of concept maps. According to them, constructed concept maps can be useful for researchers who are beginners to the field of e-learning, for teachers to develop adaptive learning materials, and for students to understand the whole picture of e-learning domain. The authors introduce the notion "relation strength" with the help of which they describe pairwise relations between extracted concepts. Relation strength is calculated on the basis of distance between two concepts in the text and on the basis their co-occurrence in one sentence. Authors link pairs of concepts which have "relation strength" more than 0.4. Like the authors of [6], the authors of this work do not label found links (do not sign them).

Generally speaking, labeling of relations extracted from the texts is a very complex problem that requires performing semantic analysis of texts. That is why many researchers note the limitedness of statistical approaches and try to combine statistical and linguistic tools by using knowledge bases suitable for semantic analysis. For example, the authors of [8] use thesaurus WordNet for part-of-speech analysis of texts. Due to determination of parts of speech in sentences they extract a predicate (the main verb) from each sentence and form for each predicate a triplet "subject-predicate-object". The subject and object are interpreted as concepts and the predicate as a relation between them (see Fig. 3). The authors of the paper are interested in building a concept map concerning biological kingdoms.

Fig. 3. "Subject-predicate-object" triplet used in [8].

The authors of [9] analyze the structure of sentences by constructing trees of dependences. They divide each sentence into a group of members dependent on the noun and a group of members dependent on the verb. They display verbs in the links and the nouns in concepts, as shown in Fig. 4. The final goal of the authors is to develop intelligent user interfaces to help understanding of complex project documents and contextualization of project tasks. The paper [10] describes an approach based on the use of thesaurus WordNet, too. The authors of this work use the lexical power of WordNet to provide the construction of an interactive concept maps by students. Using WordNet, the authors perform processing of different student responses revealing the meaning of the concepts with the help of synonyms hyponyms, meronyms, and homonyms existing in the lexical base of WordNet.

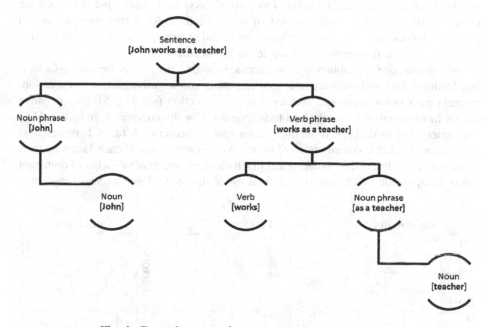

Fig. 4. Dependency tree for semantic analysis used in [9].

Like the authors of [6], the authors of [11] search for "noun-verb-noun" structures in sentences. They use verbs as designations of links and display nouns in concepts. The authors of [12] use not only verbs but also prepositional groups of the English language which designate possessiveness (of), direction (to), means (by), etc. for designation of links. The authors of [13] propose a novel approach based on combined techniques of automatic generation of exhaustive syntactic rules, restricted-context part-of-speech tagging and vector space intersection. They start from a basic set of simple syntactic rules (Noun-Verb-Noun, Verb-Noun-Verb) and expand the concept of noun (N) to include other syntagmatic components.

All the enumerated works demonstrate quite good results for extraction of concepts and relations. The problem only occurs when marking relations, i.e. when assigning semantics to relations. Interpretation of verbs and prepositional groups as relations is one of the ways to solve this problem which requires the use of linguistic tools and dictionaries.

3 Proposed Approach

3.1 Concepts Extraction

The first step of our approach is extraction of domain key terms which can be used as concepts – basic elements of a concept map. We start this step with preprocessing of a given textual collection, i.e. division texts into words, lemmatization (reduction of words to normal forms) and removal of stop-words. As result of such preprocessing we obtain a list of unique words (terms) of the collection. After that we construct a term-document matrix the rows of which correspond to terms, columns – to documents and elements – to frequencies of using terms in documents.

We decompose the obtained co-occurrence matrix into scores (terms-by-factors) and loadings (factors-by-documents) matrixes using non-negative matrix factorization, wherein each factor represents one topic of the collection (see Fig. 5). Non-negative matrix factorization is a very fruitful technique used for dimensionality reduction [14]. It produces data projections in the new factor space wherein each factor is represented as a vector of relative contributions of terms. We can sort terms of each factor by their contributions in descending order, select first p elements and generate a list of dominant terms. Conjunction of all lists gives a final list of dominant domain terms (concepts).

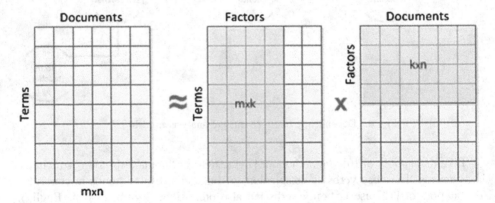

Fig. 5. Non-negative matrix factorization.

3.2 Relations Extraction

The obtained term-document matrix contains information concerning links between all terms and all documents. To extract relations between concepts, we should concentrate on links between selected key terms. So we should exclude from our term-document matrix rows which do not correspond to key terms selected on the previous step. Thereby we should reduce the dimension of our matrix. Then we should transform this reduced term-document matrix to a term-term matrix. For this, we should find pairwise distances between rows of the term-document matrix. The distance can be calculated using the cosine measure:

$$c = \cos(\bar{x}, \bar{y}) = \frac{\bar{x} \cdot \bar{y}}{|x| \cdot |y|}$$

where c is a distance value; x, y are any two rows in the reduced term-document matrix corresponding to the pair of concepts. The obtained values are measured by figures in the range from 0 to 1. The higher the similarity between vectors-terms, the less is the angle, the higher is the cosine of the angle (cosine measure). Consequently, maximum similarity is equal to 1, and minimum one is equal to 0.

The obtained term-term matrix measures distances between terms based on their co-occurrence in documents (as coordinates of vectors-terms are frequencies of their use in documents). It means that the sparser the initial term-document matrix, the worse is the quality of the term-term distances matrix. Therefore, it is expedient to save the initial matrix from information noise and rarefaction with the help of the latent semantic analysis [15]. The presence of noise is conditioned by the fact that, apart from the domain knowledge, initial documents contain "general places" which, nevertheless, contribute to the statistics of distribution.

We use the method of latent semantic analysis for clearing up the matrix from information noise. The essence of the method is based on approximation of the initial sparse and noised matrix by a matrix of lesser rank with the help of singular decomposition. Singular decomposition of matrix A with dimension $M \times N$, $M > N$ is its decomposition in the form of product of three matrices – an orthogonal matrix U with dimension $M \times M$, diagonal matrix S with dimension $M \times N$ and a transposed orthogonal matrix V with dimension $N \times N$:

$$A = USV^T \tag{1}$$

Such decomposition has the following remarkable feature. Let matrix A be given for which singular decomposition $A = USV^T$ is known and which is needed to be approximated by matrix A_k with the pre-determined rank k. If in matrix S only k greatest singular values are left and the rest are substituted by nulls, and in matrices U and V^T only k columns and k lines are left, then decomposition

$$A_k = U_k S_k V_k^T \tag{2}$$

will give the best approximation of the initial matrix A by matrix of rank k. Thus, the initial matrix A with the dimension $M \times N$ is substituted with matrices of lesser sizes $M \times k$ and $k \times N$ and a diagonal matrix of k elements. In case when k is significantly less than M and N, we have a significant compression of information. However, part of information is lost and only the most important (dominant) part is saved. The loss of information takes place because of neglecting small singular values, i.e. the more singular values are discarded the higher the loss. Thus, the initial matrix gets rid of information noise introduced by random elements.

3.3   Summarization

The extracted concepts and relations must be plotted on a concept map. Let us repeat that as concepts we use terms which contribution to collection factors is higher than a certain threshold value determined experimentally. Varying this value, we can reduce or increase the list of concepts. In the same way, we can vary the number of extracted relations. Among all pairwise distances in the term-term matrix we select the values higher than a certain threshold. Therefore, we select only edges (links) which connect only the concepts the proximity between which is higher than the indicated threshold.

4   Experiments

To carry out experiments, we chose the subject domain "Ontology engineering". The documents representing chapters from the textbook [16] formed a teaching collection. Tokenization and lemmatization from the collection resulted in a thesaurus of unique terms. The use of non-negative matrix factorization allowed selecting 500 key concepts of the subject domain. Table 1 presents the first 12 concepts.

Table 1.   Key extracted concepts.

No	Concept
1	Semantic
2	Web
3	Property
4	Manner
5	Model
6	Class
7	Major
8	Side
9	Word
10	Query
11	Rdftype
12	Relationship

Then the constructed term-document matrix was approximated by a matrix of the rank 100 with the help of singular decomposition. On the basis of the obtained matrix, pairwise distances between terms-lines were calculated using cosine measure. Thus, the transfer from a term-document matrix to a term-term matrix was carried out. Table 2 presents, as an example, some pairs of terms with different indexes of proximity. Only the links the proximity values of which exceeded 0.5 were left as relations significant for construction of a concept map.

Having obtained all concepts and links, we constructed a graph of the concept map. The concepts were taken as nodes of the graph and relations between concepts were taken as edges. As the general structure of the map is too large for analysis, we present a fragment of this map in Fig. 6.

Table 2. The samples of various extracted relations.

No	First concept	Second concept
1	OWL	Class
2	OWL	Modeling
3	OWL	Member
4	Property	Class
5	Result	Pattern
6	Term	Relationship

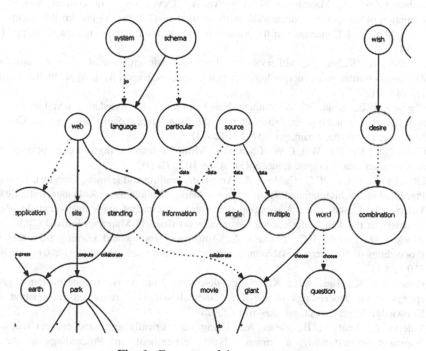

Fig. 6. Fragment of the concept map.

5 Conclusion

We are introduced another method of constructing concept maps and experimental results have been positively evaluated by two independent experts in the domain. Further studies will be related with the processing of large concept maps, their visualization and intelligent processing methods.

This work is part of a project carried out in the Al-Farabi Kazakh National University, the goal of which is to develop efficient algorithms and models of semi-structured data processing, on the basis of modern technologies in the field of the Semantic Web using the latest high-performance computing achievements to obtain new information and knowledge from unstructured sources, large amounts of scientific data and texts.

References

1. Sherman, R.: Abstraction in concept map and coupled outline knowledge representations. J. Interact. Learn. Res. **14**, 31–49 (2003)
2. Villalon, J., Calvo, R.: Concept map mining: a definition and a framework for its evaluation. In: Proceedings of the International Conference on Web Intelligence and Intelligent Agent Technology, vol. 3, Los Alamitos, USA, pp. 357–360 (2008)
3. Villalon J., Calvo R., Montenegro R.: Analysis of a gold standard for Concept Map Mining – how humans summarize text using concept maps. In: Proceedings of the Fourth International Conference on Concept Mapping, pp. 14–22 (2010)
4. Akhmed-Zaki D., Mansurova M., Pyrkova A.: Development of courses directed on formation of competences demanded on the market of IT technologies. In: Proceedings of the 2014 Zone 1 Conference of the American Society for Engineering Education, pp. 1–4 (2014)
5. Zubrinic, K., Kalpic, D., Milicevic, M.: The automatic creation of concept maps from documents written using morphologically rich languages. Expert Syst. Appl. **39**(16), 12709–12718 (2012)
6. Clariana, R.B., Koul, R.: A computer-based approach for translating text into concept map-like representations. In: Proceedings of the First International Conference on Concept Mapping, Pamplona, Spain, pp. 131–134 (2004)
7. Chen, N.S., Kinshuk Wei, C.W., Chen, H.J.: Mining e-learning domain concept map from academic articles. Comput. Educ. **50**(3), 1009–1021 (2008)
8. Oliveira, A., Pereira, F.C., Cardoso, A.: Automatic reading and learning from text. In: Paper Presented at the International Symposium on Artificial Intelligence Kolhapur, India (2001)
9. Valerio, A., Leake, D.B.: Associating documents to concept maps in context. In: Paper Presented at the Third International Conference on Concept Mapping, Finland (2008)
10. Alves, Z.O., Pereira, F.C., Cardoso, A.: Automatic reading and learning from text. In: Proceedings of the International Symposium on Artificial Intelligence (ISAI 2001), pp. 302–310 (2001)
11. Rajaraman, K., Tan, A.H.: Knowledge discovery from texts: a concept frame graph approach. In: Proceedings of the 11th International Conference on Information and Knowledge Management, pp. 669–671 (2002)
12. Valerio, A., Leake, D.B., Cañas, A.J. Using automatically generated concept maps for document understanding: a human subjects experiment. In: Proceedings of the 15 International Conference on Concept Mapping, pp. 438–445 (2012)
13. Reis, J.C., Gaia, A.S.C., Viegas Jr., R.: Concept maps construction based on exhaustive rules and vector space intersection. IJCSNS **14**(7), 26 (2014)
14. Costa, G., Ortale, R., A latent semantic approach to xml clustering by content and structure based on non-negative matrix factorization. In: 2013 12th International Conference on Machine Learning and Applications (ICMLA) IEEE 2013, vol. 1, pp. 179–184 (2013)
15. Evangelopoulos, N.E.: Latent semantic analysis. Wiley Interdisc. Rev.: Cognitive Sci. **4**(6), 683–692 (2013)
16. Allemang, D., Hendler, J.: Semantic Web for the Working Ontologist, 2nd edn. Elsevier Inc., Philadelphia (2011)

A Comparison of Term Weighting Schemes for Text Classification and Sentiment Analysis with a Supervised Variant of tf.idf

Giacomo Domeniconi, Gianluca Moro, Roberto Pasolini[✉],
and Claudio Sartori

DISI, Università degli Studi di Bologna, Bologna, Italy
{giacomo.domeniconi,gianluca.moro,roberto.pasolini,
claudio.sartori}@unibo.it

Abstract. In text analysis tasks like text classification and sentiment analysis, the careful choice of term weighting schemes can have an important impact on the effectiveness. Classic unsupervised schemes are based solely on the distribution of terms across documents, while newer supervised ones leverage the knowledge of membership of training documents to categories; these latter ones are often specifically tailored for either topic or sentiment classification. We propose here a supervised variant of the well-known tf.idf scheme, where the idf factor is computed without considering documents within the category under analysis, so that terms frequently appearing only within it are not penalized. The importance of these terms is further boosted in a second variant inspired by relevance frequency. We performed extensive experiments to compare these novel schemes to known ones, observing top performances in text categorization by topic and satisfactory results in sentiment classification.

Keywords: Term weighting · Supervised term weighting · Text classification · Sentiment analysis

1 Introduction

The classification of text documents written in natural language in a number of predefined *categories* (or *classes*) is a task performed in many practical applications. In the canonical *text classification* or *categorization* task, for which many tailored solutions exist, the categories correspond to discussion topics (such as *sports, movies, travels* etc.) and the goal is to label each document with the topic it deals with, possibly more than one. In the typical machine learning-based approach, one or more classifiers are trained from pre-labeled documents and used to automatically label subsequent documents [24]. Other tasks exist with a different premise and using different techniques with respect to understanding topics discussed in text, but are structurally ascribable as text classification tasks. An example is the identification of languages used in documents, where categories represent languages rather than topics.

© Springer International Publishing Switzerland 2016
M. Helfert et al. (Eds.): DATA 2015, CCIS 584, pp. 39–58, 2016.
DOI: 10.1007/978-3-319-30162-4_4

In text analysis, a research branch currently of high interest is *sentiment analysis*, where the opinion of people must be understood from what they write. We specifically consider the *sentiment classification* task, where documents must be classified according to the expressed attitude towards a specific subject or object, which may be a politician, a restaurant, a book etc. In a very common case documents are reviews, such as those written by users of e-commerce sites, which must be classified as either *positive* or *negative*, or even *neutral* if such case is considered. This can then be seen as another example of text classification task, where categories represent opposite polarities of judgement.

In any text analysis task, the first hurdle is the unstructured nature of text data, which requires to employ a more suitable representation of documents. The most widely used approach for this is the Vector Space Model (VSM), where each document is represented as a vector in a multi-dimensional space. Features of these vectors usually correspond to specific words or *terms*, indicating their presence within each document without considering their position or their function: such vectors are also known as *bags of words*. After extracting single terms from documents and selecting a suitable set of them to be used as features, a vector for each document must be extracted with suitable values or *weights* indicating the importance of each term in the document.

These weights are consistently assigned according to a chosen *term weighting scheme*. Although few of them are well-known and largely employed, many different schemes and variants thereof have been proposed throughout the literature, each generally carrying some kind of improvement over previous ones. Given a text analysis task where VSM is employed, the choice of the term weighting scheme can have a significant impact on its outcomes.

For example, in [18] different weighting schemes are thoroughly tested on text classification by topic with SVM classifiers and substantial gaps can be observed between accuracy levels obtained with different schemes. Similarly, more recent studies [6,22] analyzed the impact of term weighting on sentiment classification, for which some schemes have been specifically devised. Other tasks where term weighting has proven to have some influence over final results are cross-domain classification [11,12], novelty mining [27] and discovery of gene-function associations [9]. From scientific research, the use of term weighting schemes has moved to practical applications, with its employment in projects of major IT enterprises such as Yahoo! [3] and IBM [23].

A weighting scheme is often composed by a *local* and a *global* factor: while the former estimates the relevance of each term within each single document regardless of other ones, the latter indicates the importance of each term across the whole collection of documents rather than in a specific one. Once weights are computed, resulting vectors are often normalized in order to avoid giving more bias to longer documents.

Many studies on term weighting focus on the global factor, for which two major approaches can be distinguished. *Unsupervised* weighting schemes are only based on the distribution of terms across documents and can be used in virtually any text analysis task. *Supervised* schemes are instead devised for classification tasks, as also leverage the information about membership of documents to categories. Older research investigated unsupervised schemes, including the well-known and widely used *tf.idf* scheme having its origins in information retrieval. More recent research is instead mostly focused on supervised schemes, especially used in text categorization and sentiment analysis.

Following a study of existing weighting schemes, both supervised and not, we propose here a variant of the classic *tf.idf* scheme, specifically by providing a supervised version of the global *idf* factor, normally based on the assumption that terms more frequent throughout a collection have minor discriminating power and are thus less important. In a text classification setting this does not necessary hold: a term appearing frequently throughout a specific category and rarely in other ones is strongly discriminative for that category, but its standard *idf* is low compared to less frequent terms. Following this intuition, the *idf* factor in our variant is computed for each category ignoring documents labeled with it, so that the importance of a term is not penalized by appearances inside the category itself. Of this scheme we also propose a further variant where it is combined with *relevance frequency*, another supervised global weighting factor.

This paper reprises work from [13], reporting the comparison between our proposed schemes and other ones in text classification by topic and also adding an overview and an evaluation of schemes aimed to sentiment analysis.

The paper is organized as follows. Section 2 gives an overview of existing term weighting methods, with focus on global factors. In Sect. 3 we introduce and motivate our two variants of *idf*. Section 4 presents the general setup of our experimental evaluation of different term weighting schemes, whose results are reported and discussed in Sect. 5. Finally, Sect. 6 sums up the work with conclusive remarks.

2   Term Weighting in Text Classification

The problem of text categorization has been extensively investigated in the past years, considering the ever-increasing applications of this discipline, such as news or e-mail filtering and organization, indexing and semantic search of documents, sentiment analysis and opinion mining, prediction of genetic diseases etc. [24]

In the machine learning approach, a knowledge model to classify documents within a set $C = \{c_1, c_2, \ldots, c_{|C|}\}$ of categories is built upon a training set $\mathcal{D}_T = \{d_1, d_2, \ldots, d_{|\mathcal{D}_T|}\}$ of documents with a known labeling $L : \mathcal{D}_T \times C \rightarrow \{0, 1\}$ ($L(d, c) = 1$ if and only if document d is labeled with c). In order to leverage standard machine learning algorithms, documents are generally pre-processed to be represented in a Vector Space Model (VSM).

In the VSM, the content of a document d_j is represented as a vector $\mathbf{w}^j = \{w_1^j, w_2^j, \ldots, w_n^j\}$ in a n-dimensional vector space $\mathbb{R}^n$, where w_i^j is a weight that indicates the importance of a term t_i in d_j. Terms $t_1, t_2, \ldots, t_n$ constitute a set of features, shared across all documents. In other words, each weight w_i^j indicates how much the term t_i contributes to the semantic content of d_j.

Weights for each term-document couple are assigned according to a predefined term weighting scheme, which must meaningfully estimate the importance of each term within each document.

Three are the considerations discussed in the years regarding the correct assignment of weights in text categorization [4]:

1. the multiple occurrence of a term in a document appears to be related to the content of the document itself (*term frequency* factor);
2. terms uncommon throughout a collection better discriminate the content of the documents (*collection frequency* factor);
3. long documents are not more important than the short ones, normalization is used to equalize the length of documents.

Following these points, most weighting schemes are the product of a *local* (*term frequency*) factor L computed for each term-document couple and a *global* (*collection frequency*) factor G computed for each term on the whole collection. *Cosine normalization* is then usually applied to each document vector.

$$\mathbf{w}^j_{normalized} = \frac{1}{\sqrt{\sum_{i=1}^n (w_i^j)^2}} \cdot \mathbf{w}^j \tag{1}$$

There are a number of ways to calculate the local term frequency factor. The simplest one is binary weighting, which only considers the presence (1) or absence (0) of a term in a document, ignoring its frequency. Another obvious possibility is to consider the number of occurrences of the term in the document, which is often the intended meaning of "term frequency" (*tf*). Among other variants, the *logarithmic tf*, computed as $\log(1+tf)$, is now practically the standard local factor used in literature [4]. In this work, we have chosen logarithmic term frequency ($\log(1 + tf)$) as the local factor for all experiments.

As mentioned earlier, the global collection frequency factor can be *supervised* or *unsupervised*, depending whether it leverages or not the knowledge of membership of documents to categories. In the following, we summarize some of the most used and recent methods proposed in the literature of both types.

2.1   Unsupervised Term Weighting Methods

Generally, unsupervised term weighting schemes, not considering category labels of documents, derive from IR research. The most widely used unsupervised method is *tf.idf*, which perfectly embodies the three assumptions previously seen. The basic idea is that terms appearing in many documents are not good for discrimination, and therefore they will weight less than terms occurring in few documents. Over the years, researchers have proposed several variations in the way they calculate and combine the three components: *tf*, *idf* and normalization.

Indicating with $N = |\mathcal{D}_T|$ the total number of training documents and with $df(t_i)$ the count of training documents where term t_i appears at least once, the standard *tf.idf* formulation is

$$tf.idf(t_i, d_j) = tf(t_i, d_j) \cdot idf(t_i) = tf(t_i, d_j) \cdot \log\left(\frac{N}{df(t_i)}\right), \tag{2}$$

where any one of the aforementioned local weighting factors can be used in place of *tf*. The *idf* factor multiplies the *tf* by a value that is greater when the term is rare in the collection $\mathcal{D}_T$ of training documents. The weights obtained by the formula above are then normalized according to the third assumption by means of cosine normalization (Eq. 1).

The standard idf factor given above can be replaced with other ones: a classic alternative is the *probabilistic idf*, defined as

$$pidf(t_i) = \log\left(\frac{N - df(t_i)}{df(t_i)}\right). \tag{3}$$

In [26] is proposed a variant of the *idf* called *Weighted Inverse Document Frequency* (*widf*), given by dividing the $tf(t_i, d_j)$ by the sum of all the frequencies of t_i in all the documents of the collection:

$$widf(t_i) = \frac{1}{\sum_{d_x \in \mathcal{D}_T} tf(t_i, d_x)} \tag{4}$$

[5] propose a combination of *idf* and *widf*, called *Modified Inverse Document Frequency* (*midf*) that is defined as follows:

$$midf(t_i) = \frac{df(t_i)}{\sum_{d_x \in \mathcal{D}_T} tf(t_i, d_x)} \tag{5}$$

Of course the simplest choice, sometimes used, is to not use a global factor at all, setting it to 1 for all terms and only considering term frequency.

2.2 Supervised Term Weighting Methods

Since text classification is a supervised learning task, where the knowledge of category labels of training documents is necessary, many term weighting methods use this information to supervise the assignment of weights to each term.

A basic example of supervised global factor is *inverse category frequency*:

$$icf(t_i) = \log\left(\frac{|\mathcal{C}|}{cf(t_i)}\right) \tag{6}$$

where $df(t_i)$ denotes the count of categories where t_i appears in at least one relevant document. The idea of the *icf* factor is similar to that of *idf*, but using the categories instead of the documents: the fewer are the categories in which a term occurs, the greater is the discriminating power of the term.

Within text categorization, especially in the multi-label case where each document can be labeled with an arbitrary number of categories, it is common to train one binary classifier for each one of the possible categories. For each category c_k, the corresponding model must separate its *positive examples*, i.e. documents actually labeled with c_k, from all other documents, the *negative examples*. In this case, it is allowed to compute for each term t_i a distinct collection frequency factor for each category c_k, used to represent documents in the VSM only for verifying relevance to that specific category.

To summarize the various methods of supervised term weighting, we show in Table 1 the fundamental elements mostly used in the following formulas to compute the global importance of a term t_i for a category c_k.

- A denotes the number of documents belonging to category c_k where the term t_i occurs at least once;
- B denotes the number of documents belonging to c_k where t_i does not occur;
- dually C denotes the number of documents not belonging to category c_k where the term t_i occurs at least once;
- finally D denotes the number of documents not belonging to c_k where t_i does not occur.

These four elements sum to N, the total number of training documents.

Table 1. Fundamental elements of supervised term weighting.

	c_k	$\overline{c_k}$
t_i	A	C
$\overline{t_i}$	B	D

The standard *idf* factor can be expressed in this notation as

$$idf = \log\left(\frac{N}{A+C}\right). \tag{7}$$

As suggested in [4,7], an intuitive approach to supervised term weighting is to employ common techniques for feature selection, such as χ^2, *information gain*, *odds ratio* and so on. In [7] the χ^2 factor is used to weigh terms, replacing the *idf* factor, and the results show that the *tf.χ^2* scheme is more effective than *tf.idf* using a SVM classifier. Similarly [4] apply feature selection schemes multiplied by the *tf* factor, by calling them "supervised term weighting". In this work they use the same scheme for feature selection and term weighting, in contrast to [7] where different measures are used. The results of the two however are in contradiction: [4] shows that the *tf.idf* always outperforms χ^2, and in general the supervised methods not give substantial improvements compared to unsupervised *tf.idf*. The widely-used collection frequency factors χ^2, information gain (ig), odds ratio (or) and mutual information (mi) are described as follows:

$$\chi^2 = N \cdot \frac{(A \cdot D - B \cdot C)^2}{(A+C) \cdot (B+D) \cdot (A+B) \cdot (C+D)}, \tag{8}$$

$$ig = -\frac{A+B}{N} \cdot \log\frac{A+B}{N} + \frac{A}{N} \cdot \log\frac{A}{A+C} + \frac{B}{N} \cdot \log\frac{B}{B+D}, \tag{9}$$

$$or = \log\left(\frac{A \cdot D}{B \cdot C}\right), \tag{10}$$

$$mi = \log \left(\frac{A \cdot N}{(A + B) \cdot (A + C)} \right). \qquad (11)$$

Any supervised feature selection scheme can be used for the term weighting. For example, the *gss* extension of the χ^2 proposed by [15] eliminates N at numerator and the emphasis to rare features and categories at the denominator.

$$gss = \frac{A \cdot D - B \cdot C}{N^2} \qquad (12)$$

Relevance frequency [17] considers the terms distribution in the positive and negative examples, stating that, in multi-label text categorization, the higher the concentration of high-frequency terms in the positive examples than in the negative ones, the greater the contribution to categorization.

$$rf = \log \left(2 + \frac{A}{\max(1, C)} \right) \qquad (13)$$

The *icf-based* scheme [28] combines this idea with *icf*:

$$\textit{icf-based} = \log \left(2 + \frac{A}{\max(1, C)} \cdot \frac{|C|}{cf(t_i)} \right) \qquad (14)$$

Further term weighting schemes are based on additional information. [25] proposes a scheme that leverages availability of past retrieval results, consisting of queries that contain a particular term, retrieved documents, and their relevance judgments. Another different approach to supervised term weighting [20] does not use the statistical information of terms in documents like methods mentioned above, but exploits instead the semantics of categories and terms.

2.3   Term Weighting Methods for Sentiment Analysis

While the sentiment classification is structurally equivalent to canonical text categorization with review polarities in place of topics, many techniques have been specifically devised for this problem, mainly in order to find and value terms expressing positive or negative sentiment. Some research on this task is focused the term weighting issue, with both studies of existing solutions and proposals of new tailored schemes, mostly supervised.

A first example of weighting scheme conceived for sentiment analysis is *delta tf.idf*, which is obtained by computing the standard idf factor separately on positive- and negative-labeled documents and taking the difference between them [21]. While this scheme is supervised as it leverages knowledge of labeling of training documents, it does not assign distinct weights to the two categories. Denoting with N^+ and df^+ total and per-term document counts restricted to positive documents and with N^- and df^- the same for negative ones, delta idf is defined as

$$\Delta idf(t_i) = \log \left(\frac{N^+}{df^+(t_i)} \right) - \log \left(\frac{N^-}{df^-(t_i)} \right) = \log \left(\frac{N^+ \cdot df^-(t_i)}{N^- \cdot df^+(t_i)} \right). \qquad (15)$$

A subsequent study [22] focused on term weighting in sentiment analysis reprised both the delta tf.idf idea and the classic BM25 scheme from information retrieval, proposing some variations of the *delta idf* factor based on probabilistic *idf* and BM25, an example of such variants is *smoothed delta idf*:

$$\Delta idf_{sm}(t_i) = \log \left(\frac{N^+ \cdot df^-(t_i) + 0.5}{N^- \cdot df^+(t_i) + 0.5} \right). \tag{16}$$

A more recent study [6] takes into consideration other supervised weighting schemes: here the unique global factor of each term, called *importance of a term for expressing sentiment* (ITS), is the maximum between the positive and negative classes of a per-class weighting scheme, such as those proposed in [4,7]. Among other schemes cited in the study is *Weighted Log Likelihood Ratio*, whose per-class value can be computed as

$$wllr(t_i, c_k) = \frac{A}{A+B} \cdot \log \left(\frac{A(C+D)}{C(A+B)} \right). \tag{17}$$

Many of the same schemes are reviewed in [14] and tested with multiple learning algorithms. Here are also proposed more trivial schemes which are shown to perform well, such as the maximum *class density*, defined as

$$cd_{MAX}(t_i) = \max_{c_k \in \mathcal{C}} cd(t_i, c_k), \quad \text{where } cd(t_i, c_k) = \frac{A}{A+C}. \tag{18}$$

In [29] is addressed the so-called *over-weighting* problem, where the weighting scheme erroneously attributes excessive importance to rare (*singular*) terms: a regularization method applicable to existing schemes is proposed as a solution.

3 A Supervised Variant of Inverse Document Frequency

Here we introduce a supervised variant of *tf.idf*. The basic idea of our proposal is to avoid decreasing the weight of terms contained in documents belonging to the same category, so that words that appear in several documents of the same category are not disadvantaged, as instead happens in the standard formulation of *idf*. We refer to this variant with the name *idfec* (Inverse Document Frequency Excluding Category). Therefore, the proposed category frequency factor scheme is formulated as

$$idfec\,(t_i, c_k) = \log \left(\frac{|\mathcal{D}_T \backslash c_k| + 1}{\max\left(1, |d \in \mathcal{D}_T \backslash c_k : t_i \in d|\right)} \right), \tag{19}$$

where "$\mathcal{D}_T \backslash c_k$" denotes training documents not labeled with c_k. Using the previous notation, the formula becomes:

$$tf.idfec\,(t_i, d_j) = tf(t_i, d_j) \cdot \log \left(\frac{C+D}{\max\left(1, C\right)} \right). \tag{20}$$

Note that with this variant of *idf* we can have particular cases. If the i-th word is only contained in j-th document, or only in documents belonging to c_k, the denominator becomes 0. To prevent division by 0, the denominator is replaced by 1 in this particular case.

The *tf.idfec* scheme is expected to improve classification effectiveness over *tf.idf* because it discriminates where each term appears. For any category c_k, the importance of a term appearing in many documents outside of it is penalized as in *tf.idf*. On the other side, the importance is not reduced by appearances in the positive examples of c_k, so that any term appearing mostly within the category itself retains an high global weight.

This scheme is similar to *tf.rf* [17], as both penalize weights of a term t_i according to the number of negative examples where the t_i appears. The difference is in the numerator of the fraction, which values positive examples with the term in *rf* and negative ones without it in *idfec*.

To illustrate these properties we use a numerical example. Considering the notation shown in Table 1, suppose we have a corpus of 100 training documents divided as shown in Table 2, for two terms t_1 and t_2 and a category c_k.

Table 2. Example of document distribution for two terms.

	c_k	$\overline{c_k}$		c_k	$\overline{c_k}$
t_1	27	5	t_2	10	25
$\overline{t_1}$	3	65	$\overline{t_2}$	20	45

We can easily note how the term t_1 is very representative, and then discriminant, for the category c_k since it is very frequent within it $(A/(A+B) = 27/30)$ and not in the rest of the documents $(C/(C+D) = 5/70)$. Similarly we can see that t_2 does not seem to be a particularly discriminating term for c_k. In the standard formulation, the *idf* is

$$idf(t_1) = \log(100/(27+5)) = \log(3.125)$$

and for our best competitor *rf* is

$$rf(t_1, c_k) = \log(2 + 27/5) = \log(7.4),$$

while with the *idfec* we obtain

$$idfec\ (t_1, c_k) = \log((65+5)/5) = \log(14).$$

For t_2 we have instead:

$$idf(t_2) = \log(2.857), \quad rf(t_2, c_k) = \log(2.4), \quad idfec(t_2, c_k) = \log(2.8).$$

We can see that our supervised version of *idf* can separate the weights of the two terms according to the frequency of terms in documents belonging to

c_k or not. In fact, while with the standard *idf* the weights of t_1 and t_2 are very similar, with *idfec* t_1 has a weight much greater than t_2 since t_1 is more frequent and discriminative for the category c_k. This kind of behavior is also exhibited by *rf*, but our method yields an even higher weight for the relevant term t_1.

In its base version, *tf.idfec* takes into account only the negative examples (C and D in Table 1). Instead it could be helpful, especially for the classification task, also to take into account how many documents belonging to c_k contain the term, i.e. how much the term occurs within the category more than in the rest of the collection. Considering this, in a way similar to [28], we propose to mix our idea with that of the *rf* in a new version of our weighting scheme, called *tf.idfec-based* (*tf.idfec-b.* for short) and expressed by the following formula:

$$tf.idfec\text{-}b.(t_i, d_j) = tf(t_i, d_j) \cdot \log\left(2 + \frac{A + C + D}{\max(1, C)}\right) \tag{21}$$

Using the example in the Table 2, the new term weighting scheme becomes for t_1 and t_2 respectively:

$$idfec\text{-}b.(t_1, c_k) = \log(21.4), \quad idfec\text{-}b.(t_2, c_k) = \log(5.2).$$

With this term weighting scheme, the difference in weight between a very common term (t_2) and a very discriminative one (t_1) is even more pronounced.

4 Experimental Setup

We performed extensive experimental evaluation to compare the effectiveness of the proposed term weighting approach with other schemes. In the following, we describe in detail the organization of these experiments.

4.1 Benchmark Datasets

We used commonly employed text collections as benchmarks for text categorization by topic and sentiment classification in our experiments.

The **Reuters-21578** corpus[1] consists in 21,578 articles collected from Reuters. According to the ModApté split, 9,100 news stories are used: 6,532 for the training set and 2,568 for the test set. One intrinsic problem of the Reuters corpus is the skewed category distribution. In the top 52 categories, the two most common categories (*earn* and *acq*) contain, respectively, 43 % and 25 % of the documents in the dataset, while the average document frequency of all categories is less than 2 %. In literature, this dataset is used considering a various number of categories: we considered two views of this corpus, *Reuters-10* and *Reuters-52* where only the 10 and the 52 most frequent categories are considered, respectively.

---

[1] http://www.daviddlewis.com/resources/testcollections/reuters21578/.

The **20 Newsgroups** corpus[2] is a collection of 18,828 Usenet posts partitioned across 20 discussion groups. Some newsgroups are very closely related to each other (e.g. comp.sys.ibm.pc. hardware/comp.sys.mac.hardware), while others are highly unrelated (e.g. misc.forsale/soc.religion.christian). Likely to [17] we randomly selected 60 % of documents as training instances and the remaining 40 % make up the test set. Contrarily to Reuters, documents of 20 Newsgroups are distributed rather uniformly across categories.

Movie Review Data[3] contains 2,000 user reviews about movies evenly distributed between positive and negative. We took the v2.0 distribution, using the first 8 of the 10 provided folds for training and the remaining 2 to test.

The **Multi-Domain Sentiment Dataset**[4] [1] contains user reviews from Amazon.com of products of different categories (*domains*): for this it is mostly used to test transfer learning methods [10]. We considered separately the domains *books*, *dvd*, *electronics* and *kitchen*, using for each the standard 2,000 reviews evenly divided between positive and negative polarity. We used the first 800 reviews for each polarity as training set and the last 200 as test set.

4.2   Documents Processing and Learning Workflow

For each dataset, all documents were pre-processed by removing punctuation, numbers and stopwords from a predefined list, then by applying the common Porter stemmer to remaining words. This does not hold for the MDSD datasets, for which we used an already pre-processed distribution where n-grams are considered as terms in addition to single words.

We performed feature selection to keep only a useful subset of terms. Specifically, we extracted for each category the p terms appearing in most of its documents, where for p the following values were tested: 25, 50, 75, 150, 300, 600, 900, 1200, 1800, 2400, 4800, 9600, 14400, 19200 (the last two only for sentiment classification). This feature selection method may be considered counterproductive since we selected the most common terms, but it is actually correct considering the use of the VSM as the terms result to be as less scattered as possible. The task of term weighting is therefore crucial to increase the categorization effectiveness, giving a weight to each term according to the category to which the documents belong.

Since we tested both supervised and unsupervised term weighting methods, we used two different procedures. For unsupervised methods we processed the training set in order to calculate the collection frequency factor for each term, which was then multiplied by the logarithmic term frequency factor (referred to as *tf* in the following) for each term in training and test set. Finally, cosine normalization (Eq. 1) was applied to normalize the term weights.

For supervised methods we used the multi-label categorization approach, where a binary classifier is created for each category. That is, for each category

---

[2] http://people.csail.mit.edu/jrennie/20Newsgroups/.

[3] http://www.cs.cornell.edu/people/pabo/movie-review-data/.

[4] https://www.cs.jhu.edu/~mdredze/datasets/sentiment/.

c_k, training documents labeled with it are tagged as positive examples, while the remaining one constitute negative examples. We computed statistical information related to c_k (as described in Table 1) for each term of training documents. Logarithmic *tf* and cosine normalization are applied as above.

To train classifiers, we chose to use support vector machines (SVM), which are usually the best learning approach in text categorization [17,24]: we tested both the linear kernel and the radial basis function (RBF) kernel. Furthermore, to test the effectiveness of classification by varying the term weighting scheme on another algorithm, we used the *Random Forest* learner [2], chosen for both its effectiveness and its speed.

4.3   Performance Evaluation

When dealing with text classification datasets, we measured the effectiveness in terms of precision (π) and recall (ρ) [19]. As a measure of effectiveness that combines π and ρ we used the well-known F_1 measure, defined as:

$$F_1 = \frac{2 \cdot \pi \cdot \rho}{\pi + \rho}. \tag{22}$$

For multi-label problems, the F_1 is estimated by its *micro-average* across categories, extracted from the component-wise sum of their confusion matrices [24].

For two-classes sentiment classification problems, we simply computed accuracy as the ratio of correctly classified test documents with respect to the total, which is equivalent to the micro-averaged F_1 measure on the two classes.

To evaluate the difference of performances between term weighting methods, we employed the McNemar's significance test [8,16,17], used to compare the distribution of counts expected under the null hypothesis to the observed counts.

Let's consider two classifiers f_a and f_b trained from the same documents but with two different term weighting methods and evaluated using the same test set: some test instances are correctly classify by both, while other ones are misclassified by one or both of them. We denote with n_{01} the number of test instances misclassified by f_a but not by f_b and vice versa with n_{10} misclassified only by f_b. The null hypothesis for the McNemar's significance test states that on the same test instances, two classifiers f_a and f_b will have the same prediction errors, which means that $n_{01} = n_{10} = 0$. So the χ statistic is defined as:

$$\chi = \frac{(|n_{01} - n_{10}| - 1)^2}{n_{01} + n_{10}} \tag{23}$$

χ is approximately distributed as a χ^2 distribution with 1 degree of freedom, where the significance levels 0.01 and 0.001 correspond respectively to the thresholds $\chi_0 = 6.64$ and $\chi_1 = 10.83$. If the null hypothesis is correct, than the probability that χ is greater than 6.64 is less than 0.01, and similarly 0.001 for a value greater than 10.83. Otherwise we may reject the null hypothesis and assume that f_a and f_b have different performances, so the two weighting schemes have a different impact when used on the particular training set.

5   Experimental Results

We tested the effectiveness of classification varying the term weighting scheme on the datasets cited above: for each one we tested the classification varying the number of features p selected for each category. Due to space constraints, we show a selection of significant numeric results, describing similarities between them and other results not shown in detail.

5.1   Text Classification by Topic

We first performed tests on topic classification datasets, which are the primary target of our weighting schemes.

Table 3 shows the performance of 11 different term weighting methods: $tf.idfec$, $tf.idfec\text{-}based$, $tf.rf$, $tf.icf\text{-}based$, $tf.idf$, $tf.\chi^2$, $tf.gss$, $tf.ig$, $midf$, $tf.oddsR$ and tf. The best micro-averaged F_1 measure across different values of the p parameter for every dataset and learning algorithm are reported. In general, our second proposed scheme $tf.idfec\text{-}based$ achieved top results in all datasets and with each classifier.

On *Reuters-52* $tf.idfec\text{-}based$ outperforms every other scheme with all classifiers: using a SVM with linear kernel, compared with the standard $tf.idf$ we have a 0.8 % improvement; the gap with standard supervised schemes is even higher. $tf.idfec\text{-}based$ also slightly outperforms $tf.rf$ and $tf.icf\text{-}based$. On *Reuters-10* the results for our proposed methods are very close to other schemes. Considering only 10 categories, supervised term weighting methods appear less relevant to classification effectiveness: this can be deduced from the difference between standard $tf.idf$ and our supervised versions on *Reuters-52*, which contains the same documents of *Reuters-10* but labeled with more categories. However, our schemes outperform standard supervised term weighting by more than 10 %. In *20 Newsgroups* $tf.idfec\text{-}based$ obtains top performances in parity with other weighting schemes. Using linear SVM, the best result of $tf.idfec\text{-}based$ is higher by 1.8 % compared to that obtained from $tf.idf$ and by 23.9 % compared with a standard supervised method like $tf.ig$.

Observing all the results, we can see that our first proposed scheme $tf.idfec$ obtains results always in line but slightly lower than the $tf.idfec\text{-}based$ variant. This evidently means that considering only the information about the negative categories of the terms is not enough to achieve an optimal accuracy. Conversely, adding information about the ratio between A and C (from notation of Table 1), it is obtained an optimal mixture that leads to better classification results, using either SVM or RandomForest classifiers.

We employ the McNemar's significance test to verify the statistical difference between performances of the term weighting schemes. We report these results in Table 4, where each column is related to a dataset and a classifier and the weighting schemes are shown in decreasing order of effectiveness. Schemes not separated with lines do not give significantly different results, a single line denotes that the schemes above perform better than the schemes below with a significance level between 0.01 and 0.001 $(A > B)$, while a double line denotes a significance level

Table 3. Micro-averaged F_1 (in thousandths) best results obtained with different global weighting factors ("b." is short for "based") on topic classification with different learning algorithms. The best result for each dataset and algorithm is marked in bold.

global wt. →	idfec	idfec-b.	rf	icf-b.	idf	χ^2	gss	ig	midf	oddsR	None
Reuters-10											
SVM(LIN)	.929	**.933**	**.933**	.930	.930	.847	.863	.852	.920	.918	.932
SVM(RBF)	.932	**.937**	**.937**	.935	.933	.879	.895	.882	.923	.926	.934
RandomForest	**.904**	.902	.903	.899	.903	.901	.901	.903	.898	.903	.902
Reuters-52											
SVM(LIN)	.920	**.925**	.922	.916	.917	.828	.848	.822	.882	.890	.912
SVM(RBF)	.925	**.927**	.926	.924	.922	.848	.873	.848	.886	.895	.915
RandomForest	.855	**.868**	.867	.853	.858	.863	.861	.864	.858	.863	.866
20 Newsgroups											
SVM(LIN)	.754	**.759**	**.759**	.747	.741	.567	.512	.520	.606	.666	.709
SVM(RBF)	.712	**.713**	.712	**.713**	.702	.587	.555	.56	.609	.664	.677
RandomForest	.536	**.570**	.563	.537	.543	.568	**.570**	.565	.536	.562	.566

better than 0.001 ($A >> B$). Results on the *Reuters-10* corpus with Random Forest are missing because there are no significant statistical differences between them. The table shows that our proposed *tf.idfec-based* scheme always provides top effectiveness and that with SVM classifiers, either linear or RBF kernel, some term weighting methods are more efficient than others. The best methods in general seem to be the latest supervised methods, such as *tf.idfec-based*, *tf.rf*, *tf.icf-based* and *tf.idfec*. Instead, using RandomForest, the classic supervised methods seem to work better, with results comparable or slightly below with respect to *tf.idfec-based*.

Let's now observe how classification effectiveness varies by changing the number of features considered to create the dictionary of the VSM. We show results on Reuters-10 and 20 Newsgroups using a linear SVM classifier in Fig. 1; results on Reuters-52 mostly follow the same trends discussed below for Reuters-10. To keep the plots readable, we show the results on a selection of schemes with best performances: *tf.idfec*, *tf.idfec-based*, *tf.rf*, *tf.icf-based*, *tf.idf*.

The plot for Reuters-10 shows that, when using *tf.idfec-based* we obtain the best results by using few features per category, considering the variations of p described in Sect. 4.2. We note that the best results on *Reuters-10* are obtained with 150 features per category and on *Reuters-52* (not shown) with 75 features, corresponding respectively to overall dictionary sizes of 498 and 970. However, we can see as the effectiveness of the classification using *tf.idfec-based* deteriorates by increasing the number of features considered, therefore introducing terms less frequent and discriminative. Analysing the behavior of the schemes from which *idfec-based* takes its cue, i.e. standard *tf.idf* and *tf.rf*, we note that this performance degradation is probably due to the *idf* factor of the weight, as even

Table 4. McNemar's significance test results. Each column is related to a dataset and a supervised classifier ('L' stands for linear kernel, 'R' for RBF), the global weighting factors are shown in decreasing order of effectiveness, separating groups of schemes by significance of differences in their performances. Results for RandomForest on *Reuters-10* are omitted as no significant difference is observed between them.

| Reuters-10 | | Reuters-52 | | | 20 Newsgroups | | |
SVM(L)	SVM(R)	SVM(L)	SVM(R)	RndFor.	SVM(L)	SVM(R)	RndFor.
idfec-b.	*idfec-b.*	*idfec-b.*	*idfec-b.*	*idfec-b.*	*idfec-b.*	*idfec-b.*	*idfec-b.*
rf	*rf*	*rf*	*rf*	*rf*	*rf*	*icf-b.*	*gss*
1	*icf-b.*	*idfec*	*idfec*	1	*idfec*	*idfec*	χ^2
icf-b.	1	*idf*	*icf-b.*	*ig*	*icf-based*	*rf*	1
idf	*idf*	*icf-b.*	*idf*	*oddsR*	*idf*	*idf*	*ig*
idfec	*idfec*	1	1	χ^2	1	1	*rf*
midf	*oddsR*	*oddsR*	*oddsR*	*gss*	*oddsR*	*oddsR*	*oddsR*
oddsR	*midf*	*midf*	*midf*	*idf*	*midf*	*midf*	*idf*
gss	*gss*	*gss*	*gss*	*midf*	χ^2	χ^2	*icf-b.*
ig	*ig*	*ig*	χ^2	*idfec*	*ig*	*ig*	*idfec*
χ^2	χ^2	χ^2	*ig*	*icf-b.*	*gss*	*gss*	*midf*

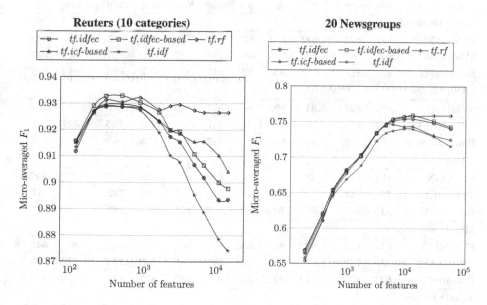

Fig. 1. Results obtained on topic classification datasets varying the number of top p features per category using a SVM classifier with linear kernel. The X axis (in logarithmic scale) indicates the resulting total number of features.

the *tf.idf* has the same type of trend results, while *tf.rf* seems to remain stable at values comparable to the best results also by increasing the dictionary size.

The right plot of Fig. 1 shows that to obtain the best results with the *20 Newsgroups* corpus is necessary a greater number of features. Using *tf.idfec-based* we obtain the best result with a dictionary of about 10000 features; after that the efficiency shows a slight decrease, but more moderate than that shown in the Reuters dataset.

5.2 Sentiment Classification

We conceived our supervised weighting schemes for text classification by topic, achieving optimal results in it. However, we tested them also on the sentiment classification, checking whether they still guarantees good accuracy levels.

In addition to those considered above, we included in the new tests some weighting schemes especially tailored for sentiment classification: *delta idf* and variants thereof, *wllr* and cd_{MAX}. As only two classes are considered (positive and negative reviews), we used higher values for the number p of features per

Table 5. Top accuracies (in thousandths) obtained with different global weighting factors ("b." is short for "based") on sentiment classification datasets with different learning algorithms. The best result for each dataset and algorithm is marked in bold.

global wt. →	idfec	idfec-b.	rf	idf	ig	midf	None	Δidf	Δidf_{sm}	wllr	cd_M
Movie Review Data											
SVM(LIN)	.842	**.845**	.842	.842	.837	.83	.837	.797	.827	.79	**.845**
SVM(RBF)	.802	.797	.797	.795	.787	.795	.787	**.807**	**.807**	.802	.792
RandomForest	.81	.815	.817	**.835**	.82	.817	.82	.807	**.835**	.812	.805
MDSD *books*											
SVM(LIN)	.802	.802	**.81**	.802	.805	.802	.805	.77	.797	.767	.8
SVM(RBF)	.742	.76	.755	.745	.745	.747	.745	**.765**	**.765**	.727	.75
RandomForest	.815	.812	**.827**	.807	.807	.817	.817	.755	.78	.745	.8
MDSD *dvd*											
SVM(LIN)	.815	.817	.83	.81	.815	.812	.815	.792	.807	.802	**.825**
SVM(RBF)	**.75**	.737	.735	.725	.715	.712	.715	.722	.722	.71	.717
RandomForest	**.812**	.807	.82	**.812**	.807	.807	.805	.77	.785	.77	.817
MDSD *electronics*											
SVM(LIN)	.842	.835	**.847**	.84	.845	.837	.845	.822	.842	.81	.842
SVM(RBF)	.772	.767	.78	.765	.762	.772	.762	**.787**	**.787**	.755	.765
RandomForest	**.847**	.825	.842	.82	.83	.825	.832	.817	.832	.832	.835
MDSD *kitchen*											
SVM(LIN)	.887	.887	**.89**	.882	.887	.885	.887	.837	.865	.857	.885
SVM(RBF)	.82	.817	.815	.81	.81	.81	.81	**.822**	**.822**	.81	.815
RandomForest	.857	.862	.867	.857	.86	.862	**.882**	.84	.847	.825	.872

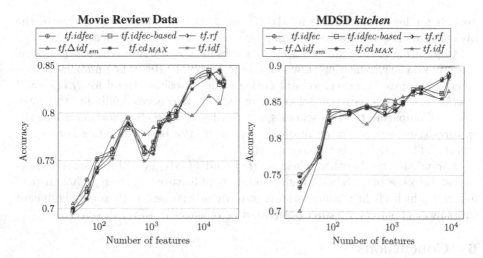

Fig. 2. Results obtained on sentiment classification datasets varying the number of top p features per category using a SVM classifier with linear kernel. The X axis (in logarithmic scale) indicates the resulting total number of features.

class, up to 19,200, as the total number of resulting features is generally lower for equal values of p. Always given the two-class setting, we evaluate here the accuracy of a classifier simply as the ratio of correctly classified test documents, as discussed above.

In Table 5, likely to above, we report the best accuracy obtained testing different values of p on the tested schemes: to give space to sentiment classification-specific schemes we removed some of the previous ones whose results were not among the best. In these tests, our schemes outperformed other ones only in some cases, with *tf.rf* and especially new schemes yielding strong performances.

Should be noted however that the differences between performances in these tests are smaller than those observed in topic classification. While some schemes generally perform better than others, the gaps are not significant in most cases. McNemar's significance test confirms this observation: also due to the relatively low number of available test documents, the performance difference between almost every possible couple of weighting schemes is statistically not significant.

The plots in Fig. 2 show how the accuracy on the Movie Review data and the MDSD *kitchen* domain using a SVM classifier with linear kernel varies for the tested values of p on a selection of the best performing weighting schemes: *tf.idf*, *tf.rf*, $tf.\Delta idf_{sm}$, $tf.cd_{MAX}$, *tf.idfec* and *tf.idfec-based*. The trends for other tested domains of MDSD are similar enough to those for *kitchen*, with small differences mostly on the overall average accuracy level.

With respect to classification by topic, we see again that the difference of performance between different weighting schemes is more restrained; this holds also for most of the schemes not shown in the plots. The number of features per class has instead an important impact on accuracy, with an especially fast growth for lower values of p in MDSD: as discussed above, this is due to the presence of only two classes, yielding an overall low number of features.

Comparing the accuracy trends as p varies for the different schemes, as above, there is no clear indication of which one is better for different numbers of considered features. However, we still notice a slight positive trend for *tf.idfec* and *tf.idfec-based* with narrower selections of features and occasionally in other situations. Comparing the two schemes, contrarily to topic classification where the *tf.idfec-based* variant always outperforms the *tf.idfec* base, the latter here sometimes yields a slightly better accuracy.

For what regards other schemes, *tf.rf* and $tf.cd_{MAX}$ exhibit some performance peaks, especially with higher numbers of features; $tf.\Delta idf_{sm}$ has instead a trend which slightly differentiates it from other ones, with some significant differences of accuracy, either positive or negative.

6   Conclusions

Starting from the general text classification problem, considering both categorization by topic and distinction between positive and negative sentiment, we briefly reviewed existing methods for both unsupervised and supervised and proposed a novel solution as a modification of the classic *tf.idf* scheme. We devised a base *tf.idfec* scheme where *idf* is computed without considering documents belonging to the modeled category: this prevents giving a low weight to terms largely present in it. The *tf.idfec-based* variant mixes our base scheme with the relevance frequency from *tf.rf*, thus also effectively boosting weights of terms which appear frequently in the category under analysis.

We performed extensive experimental studies on benchmark datasets for text classification by topic and by sentiment: the *Reuters* corpus with either 10 or 52 categories and *20 Newsgroups* for the former, the *Movie Review Data* and the *Multi-Domain Sentiment Dataset* for the other. We compared our two weighting schemes against other known ones using both SVM and Random Forest learning algorithms and different levels of feature selection.

The results show that the *tf.idfec-based* method combining *idfec* and *rf* generally gets top results for topic classification. Through statistical significance tests, we showed that the proposed scheme always achieves top effectiveness and is never worse than other methods. The results show a close competition between our *tf.idfec-based* and *tf.rf*: the best results obtained with the different datasets and algorithms, varying the amount of feature selection, are very similar, but with some differences. *tf-rf* seems more stable when the number of features is high, while our *tf.idfec-based* gives excellent results with few features and shows some decay (less than 4 %) when the number of features increases.

On the other side, tests on sentiment classification suggest that gaps between different weighting schemes, including some of them specifically devised for this task, are not statistically significant, although also due in our case to the low number of test documents. Here our schemes are not superior to other ones, as they have not been developed for this task, but generally appear to be nearly as good as those yielding better accuracy estimates.

In the future, we plan to test further variants of our scheme, possibly inspired by existing ones and trying to obtain significantly good results also in sentiment classification, for which we aim to perform tests on larger datasets. Also, we are considering to expand the study of existing term weighting schemes for text categorization and sentiment analysis to a more complete survey, complete of experimental comparison of their performances on several datasets.

References

1. Blitzer, J., Dredze, M., Pereira, F.: Biographies, bollywood, boom-boxes and blenders: domain adaptation for sentiment classification. Assoc. Comput. Linguist. **7**, 440–447 (2007)
2. Breiman, L.: Random forests. Mach. Learn. **45**(1), 5–32 (2001)
3. Carmel, D., Mejer, A., Pinter, Y., Szpektor, I.: Improving term weighting for community question answering search using syntactic analysis. In: Proceedings of the 23rd ACM International Conference on Conference on Information and Knowledge Management, CIKM 2014, pp. 351–360. ACM, New York (2014)
4. Debole, F., Sebastiani, F.: Supervised term weighting for automated text categorization. In: Proceedings of the 18th ACM Symposium on Applied Computing, SAC 2003, pp. 784–788. ACM Press (2003)
5. Deisy, C., Gowri, M., Baskar, S., Kalaiarasi, S., Ramraj, N.: A novel term weighting scheme midf for text categorization. J. Eng. Sci. Technol. **5**(1), 94–107 (2010)
6. Deng, Z.H., Luo, K.H., Yu, H.L.: A study of supervised term weighting scheme for sentiment analysis. Expert Syst. Appl. **41**(7), 3506–3513 (2014)
7. Deng, Z.-H., Tang, S., Yang, D., Li, M.Z.L.-Y., Xie, K.-Q.: A comparative study on feature weight in text categorization. In: Yu, J.X., Lin, X., Lu, H., Zhang, Y. (eds.) APWeb 2004. LNCS, vol. 3007, pp. 588–597. Springer, Heidelberg (2004)
8. Dietterich, T.G.: Approximate statistical tests for comparing supervised classification learning algorithms. Neural Comput. **10**(7), 1895–1923 (1998)
9. Domeniconi, G., Masseroli, M., Moro, G., Pinoli, P.: Random perturbations of term weighted gene ontology annotations for discovering gene unknown functionalities. In: Fred, A., Dietz, J.L.G., Aveiro, D., Liu, K., Filipe, J. (eds.) IC3K 2014. CCIS, vol. 553, pp. 181–197. Springer, Heidelberg (2015)
10. Domeniconi, G., Moro, G., Pagliarani, A., Pasolini, R.: Markov chain based method for in-domain and cross-domain sentiment classification. In: Proceedings of the 7th International Conference on Knowledge Discovery and Information Retrieval (2015)
11. Domeniconi, G., Moro, G., Pasolini, R., Sartori, C.: Cross-domain text classification through iterative refining of target categories representations. In: Proceedings of the 6th International Conference on Knowledge Discovery and Information Retrieval (2014)
12. Domeniconi, G., Moro, G., Pasolini, R., Sartori, C.: Iterative refining of category profiles for nearest centroid cross-domain text classification. In: Fred, A., Dietz, J.L.G., Aveiro, D., Liu, K., Filipe, J. (eds.) IC3K 2014. CCIS, vol. 553, pp. 50–67. Springer, Heidelberg (2015). doi:10.1007/978-3-319-25840-9_4
13. Domeniconi, G., Moro, G., Pasolini, R., Sartori, C.: A study on term weighting for text categorization: a novel supervised variant of tf.idf. In: 4th International Conference on Data Management Technologies and Applications (2015)

14. Fattah, M.A.: New term weighting schemes with combination of multiple classifiers for sentiment analysis. Neurocomputing **167**, 434–442 (2015)
15. Galavotti, L., Sebastiani, F., Simi, M.: Experiments on the use of feature selection and negative evidence in automated text categorization. In: Borbinha, J.L., Baker, T. (eds.) ECDL 2000. LNCS, vol. 1923, pp. 59–68. Springer, Heidelberg (2000)
16. Lan, M., Sung, S.Y., Low, H.B., Tan, C.L.: A comparative study on term weighting schemes for text categorization. In: Proceedings of the 2005 IEEE International Joint Conference on Neural Networks, IJCNN 2005, vol. 1, pp. 546–551. IEEE (2005)
17. Lan, M., Tan, C.L., Su, J., Lu, Y.: Supervised and traditional term weighting methods for automatic text categorization. IEEE Trans. Pattern Anal. Mach. Intell. **31**(4), 721–735 (2009)
18. Leopold, E., Kindermann, J.: Text categorization with support vector machines. How to represent texts in input space? Mach. Learn. **46**(1–3), 423–444 (2002)
19. Lewis, D.D.: Evaluating and optimizing autonomous text classification systems. In: Proceedings of the 18th Annual International ACM SIGIR Conference on Research and Development in Information Retrieval, SIGIR 1995, pp. 246–254. ACM, New York (1995)
20. Luo, Q., Chen, E., Xiong, H.: A semantic term weighting scheme for text categorization. Expert Syst. Appl. **38**(10), 12708–12716 (2011)
21. Martineau, J.C., Finin, T.: Delta tfidf: An improved feature space for sentiment analysis. In: Third International AAAI Conference on Weblogs and Social Media (2009)
22. Paltoglou, G., Thelwall, M.: A study of information retrieval weighting schemes for sentiment analysis. In: Proceedings of the 48th Annual Meeting of the Association for Computational Linguistics, ACL 2010, pp. 1386–1395. Association for Computational Linguistics, Stroudsburg (2010)
23. Papineni, K.: Why inverse document frequency? In: Proceedings of the Second Meeting of the North American Chapter of the Association for Computational Linguistics on Language technologies. pp. 1–8. Association for Computational Linguistics (2001)
24. Sebastiani, F.: Machine learning in automated text categorization. ACM Comput. Surv. **34**(1), 1–47 (2002)
25. Song, S.K., Myaeng, S.H.: A novel term weighting scheme based on discrimination power obtained from past retrieval results. Inf. Process. Manage. **48**(5), 919–930 (2012)
26. Tokunaga, T., Makoto, I.: Text categorization based on weighted inverse document frequency. In: Special Interest Groups and Information Process Society of Japan (SIG-IPSJ). Citeseer (1994)
27. Tsai, F.S., Kwee, A.T.: Experiments in term weighting for novelty mining. Expert Syst. Appl. **38**(11), 14094–14101 (2011)
28. Wang, D., Zhang, H.: Inverse-category-frequency based supervised term weighting schemes for text categorization. J. Inf. Sci. Eng. **29**(2), 209–225 (2013)
29. Wu, H., Gu, X.: Reducing over-weighting in supervised term weighting for sentiment analysis. In: Proceedings of the 25th International Conference on Computational Linguistics, COLING 2014 (2014)

A Framework for Spatio-Multidimensional Analysis Improved by Chorems: Application to Agricultural Data

François Johany[1(✉)] and Sandro Bimonte[2]

[1] INRA, UMR Métafort, Aubiere, France
francois.johany@clermont.inra.fr
[2] Irstea, TSCF, Aubiere, France
sandro.bimonte@irstea.fr

Abstract. Spatial OLAP (SOLAP) systems are decision-support systems for the analysis of huge volumes of spatial data. Usually, SOLAP clients provide decision-makers with a set of graphical, tabular and cartographic displays to visualize warehoused spatial data. Geovisualization methods coupled with existing SOLAP systems are limited to interactive (multi) maps. However, a new kind of geovisualization method recently appears to provide summaries of geographic phenomena: the chorem-based methods. A chorem is theoretically defined as a schematized spatial representation, which eliminates any unnecessary details to the map comprehension. Therefore, in this paper we investigate the opportunity to integrate SOLAP and chorem systems in a unique decision-support system. We propose the ChoremOLAP system that enriches SOLAP maps with chorems. We apply our proposal to agricultural data analysis, since both chorems and SOLAP have been rarely used in this application domain. Using open data provided by the FAO, we show how ChoremOLAP is well adapted in the agricultural context.

Keywords: Spatial OLAP · Geovisualization · Chorems · Spatial data warehouse

1 Introduction

Spatial On-Line Analytical Processing (SOLAP) systems allow on-line analyzing huge volume of spatial data to provide numerical indicators according to some analysis axes [1]. SOLAP has been successfully applied in several application domains such as health, agriculture, etc. SOLAP systems integrate Geographic Information Systems (GIS) functionalities with OLAP systems to provide a cartographic visualization of these indicators [2]. Decision-makers trigger SOLAP operators by the simple interaction with visual components of SOLAP clients (pivot tables, graphical and cartographic displays). Therefore, they can easily and interactively explore geo-referenced data looking for unknown and/or unexpected patterns and/or confirm their decisional hypothesis on some spatial phenomena. The success of SOLAP rests on the visual analytic paradigm "Analyze First - Show the Important - Zoom, Filter, Analyze Further - Details on Demand" [30], and its adaptation to geographic information, the so called

© Springer International Publishing Switzerland 2016
M. Helfert et al. (Eds.): DATA 2015, CCIS 584, pp. 59–80, 2016.
DOI: 10.1007/978-3-319-30162-4_5

Geovisualization. Geovisualization integrates the techniques of scientific visualization, cartography, image analysis, and data mining to provide a theory of methods and tools for the representation and discovery of spatial knowledge [27]. Geovisualization analytics tasks are performed using SOLAP operators (Slice and Dice, Roll-up and Drill-down) whose results are displayed in interactive thematic maps. However, a part from thematic maps, SOLAP systems lack of advanced Geovisualization techniques as described in [5]. In particular, summarizing information (i.e. Zoom visual analytic task) is reduced to aggregation of measures values using SQL aggregation functions of Roll-Up operator (e.g. SUM, MIN, MAX), but no additional visual summary is provided.

Consequently, sometimes SOLAP cartographic displays are not well adapted to complex spatial phenomena, which need several or temporal indicators leading to useless and/or clutterd maps [36].

Per contra, recent results have demonstrated that chorems can be used to both catch a thematic global view of a territory and its phenomena [11, 14], and investigate complex spatial phenomena by accessing data characterizing them. A chorem is a schematized spatial representation, which eliminates any detail unnecessary to the map comprehension [9]. The main limitation of these approaches is that chorem map extraction cannot be done on-demand according to spatial decision-makers needs. This limits the potentiality of the spatial decision-making process, since, as stated in, [27] high interactivity exploration and analysis are mandatory when dealing with complex and unknown datasets.

In this paper we carry out our aim of supporting users' tasks through advanced solutions and systems, and describe how chorems can be used to perform on-line analysis through a simple and immediate approach that exploits analysis capabilities of SOLAP systems. On this basis, in this paper we propose a unique framework, called ChoremOLAP, that, starting from warehoused spatial datasets, is able to on-line produce and visualize chorem maps exploiting the functionalities of SOLAP systems. We validate our approach in the context of agricultural data analysis using open data from FAO.

The paper is organized as follows. Section 2 recalls some basic requirements of agricultural data analysis; Sect. 3 describes main concepts about SOLAP systems and chorems; Related work about geovisualization in SOLAP and chorems, and the usage of those systems in the agricultural context are detailed in Sect. 4; Sect. 5 presents a real case study; The ChoremOLAP system is described in Sect. 6; Sect. 7 presents a discussion of our proposal; Conclusions and future work are drawn in Sect. 8.

2   Context of Agricultural Data

The current context of data in agriculture is marked by the Spatial Big Data era. Spatial Big Data is defined as an extension of Big Data, which is characterized by the well-known Variety, Velocity, Volume and Veracity properties, with spatial data (Shekhar et al. 2012) [35]. Spatial Big Data is mostly produced by citizens via social networks ("contributed data") and Volunteer Geographic Information (VGI) systems ("volunteer data") (Spinsanti et al. 2013) [37]. In this context, decision-makers need information systems able to analyse huge volume of spatial data.

Also, the transformation of agriculture towards agro-ecology dynamics and precision practices implies to improve the knowledge of the production process instead of the technical issues of the production. This goal can be achieved exclusively when decision-makers are able to acquire a precise and shared knowledge about agricultural data, activities, products, and their relations with other application domains (e.g. environmental data). Therefore, the challenge is to provide decision-support systems with a set of functionalities to cross different domains data and decision-makers knowledge.

Finally, it also important underline that with Spatial Big Data, citizens are not only data producers (by means of VGI tools like Openstreet map), but they also play the role of decision-makers. Indeed, only their participation in the public decision-making process could grant a social durability of political actions. Then, in this particular context, decision-makers need information systems supported with user-interfaces more usable and simple as possible.

As described in the next sections, Spatial OLAP and chorems systems seem decision-making tools supporting above described requirements for agricultural data analysis (volume, variety and easiness of use).

3 Main Concepts

3.1 Spatial OLAP

Data warehouses (DW) and OLAP systems are Business Intelligence technologies aim to support on-line analysis of huge amounts of data. Warehoused data are structured according the so-called multidimensional model, which represents data according to different analysis axes (dimensions) and facts [19]. Dimensions are composed of hierarchies, which define groups for data (members) used as analysis axes. Facts represent the subjects of analysis, and they are described by numerical measures, which are analysed at different granularities associated to the levels of hierarchies. Decision-makers can aggregate measures at coarser hierarchy levels using classical SQL aggregation functions (AVG, SUM, MIN, MAX, COUNT). OLAP operators are defined to explore warehoused data. Classical OLAP operator are: Slice which selects of a part of the data warehouse, Dice which projects a dimension, RollUp which aggregates measures climbing on a dimension hierarchy, and DrillDown, which is the reverse of RollUp.

Since OLAP systems do not allow to integrate spatial data into the analysis and exploration process, Spatial OLAP (SOLAP) systems have been introduced [1]. SOLAP systems integrate OLAP and Geographic Information System functionalities in a unique framework to take advantage from the analysis capabilities associated to spatial data. SOLAP redefines main OLAP concepts. In particular, the integration of spatial data in OLAP dimensions brings to the definition of spatial dimension: non geometric dimension, spatial geometric dimension (i.e. members with a cartographic representation) or mixed spatial dimension (i.e. combine cartographic and textual members). When the studied subject of the decision process is the spatial information itself, then the concept of spatial measures are introduced. A SOLAP system allows visualizing results of SOLAP queries using interactive tabular and map displays.

A typical SOLAP architecture is composed of a Spatial DBMS to store (spatial) data (for example Spatial Oracle, PostGIS, etc.); a SOLAP server, which implements the SOLAP operators (for example GeoMondrian, etc.); and a SOLAP client (e.g. Map4Decision), which combines and synchronizes tabular, graphical, and interactive maps to visualize the results obtained after executing SOLAP queries. In particular, data collected using several heterogeneous data sources are integrated in the SDW tier by means of Spatial Extraction, Transformation, and Loading tools (Spatial ETL). Such tools as Spatial Data Integrator, GeoKeetle), provide a set of operators to work with spatial data. Examples of such operators are rename, aggregate, etc. [4, 17, 28].

Warehoused (spatial) data are well-structured good quality data since data sources are cleaned using (Spatial) ETL tools that implement a quality control protocol.

This kind of SOLAP architecture is referred to as OLAP-GIS integrated. Indeed, three kinds of SOLAP architecture have been defined [1]. GIS dominant SOLAP tools do not present the OLAP server, and they define OLAP queries using classical SQL queries [33]. Therefore, they don't take advantage of hierarchical navigation and advanced calculation features of OLAP servers. OLAP dominant tools [5] do not provide advanced GIS functionalities allowing for only simple cartographic visualization, limiting spatial analysis capabilities. Therefore, it has been recognized that OLAP-GIS solutions such as described by [20], are best suited for an effective SOLAP analysis [2].

3.2 Chorems

In the last few years, much work has been done on the chorem concept and on its exploitation as an appealing visual notation to convey information about phenomena occurring in specific application domains, such as land use and territorial management. The term chorem derives from the Old Greek word χώρα (read chora). According to the definition of the French geographer Roger Brunet [9], a chorem is a schematized spatial representation, which eliminates any unnecessary details to the map comprehension. A chorem is a synthetic global view of data of interest which emphasizes salient aspects. Figure 1 shows an example of a chorem map, which contains chorems referring to the environmental dynamics of the area around the city of Poitiers, France.

This map results from a participatory process with agents of DREAL Poitou-Charentes (regional office of the environment) [26]. They handy draw all the information needed using a Geographic Information System. Then, they simplify the geometries of the map using the GIS functionalities according an adapted semiotic. This chorem map shows the predominance of transport corridors in the organization of the territory, the urban continuity and the expansion of the cities. Moreover, chorems show the interaction between these dynamics and the environmental issues of this area.

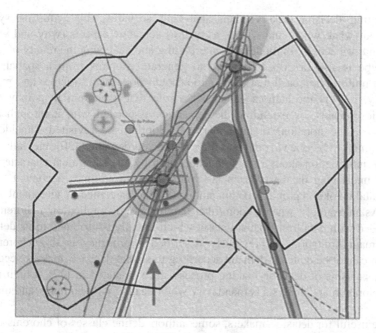

Fig. 1. Chorems example.

4 Related Work

4.1 Geovisualization

Few works investigate using advanced geovisualization methods for SOLAP. A survey can be found in [3, 5] study the usage of the space-time cube geovisualization methodology to visualize spatial measures inside pivot table cells. The integration of geobrowsers (such as Google Earth) in to OLAP clients has been successfully explored in [15]. Indeed geobrowsers allow a 3D visualization of spatial members, and adding contextual external geographic information (i.e. layers) via web services. [2] adds multimedia elements such as photos, videos, etc. to spatial data warehouse data. This approach allows contextualizing SOLAP in the sense that additional and complementary data can be used to better understand and explain SOLAP query results.

To conclude, existing SOLAP tools are limited to classical GIS visualization maps such as interactive cloropleth and thematic maps.

The evolution of chorems both in terms of applications and semantics is extensively discussed in [13] where the author provides a review about the history of chorem, from its definition [9] to recent applications [32].

A more recent application of the chorem concept has been illustrated [14] where the authors show how chorems can be used to visually summarize database content. To this aim they provide for a definition and a classification of chorems meant both to homogenize chorem construction and usage, along with a usable framework for computer systems. In particular, this work emphasizes the role that chorems may play in supporting decision-makers when analyzing scenarios, by acquiring syntactic information (what, where and when), as well as semantic aspects (why and what if), useful to human activity of modelling, interpreting and analysing the reality of interest. A prototype is also described, targeted to generate chorems from a spatial dataset through a uniform approach that takes into account both their structure and meaning.

Finally, in [11] the authors enhance the role that a chorem map may play in geographic domains, by extending the semantics associated with it through a more expressive visual notation. In particular, by adopting the revisited Shneiderman's mantra, namely "Overview, zoom and filter, details on demand" [30], they allow users to acquire information about a single phenomenon by accessing data characterizing it from the underlying database. Each task of interaction assumes a context-sensitive meaning and invokes a proper function among the ones specified in agreement with the mantra. As an example, when a zoom/filter combination is applied on a chorem, users are provided with data from spatial dataset which initially contributed to its definition.

The main limitation of previously described approaches is that chorem map extraction cannot be done on-demand according to spatial decision-makers needs. This limits the analysis of decision-making process since as stated in [27], a high interactivity exploration and analysis is mandatory when dealing with complex and unknown datasets.

To be useful for decision-makers, some authors define classes of chorems in order to help decision-makers to choice the right visual representation for a particular phenomenon. Therefore, from the initial chorematic grid of Brunet, JP Deffontaines et al. [9, 12] have formalized spatial models to represent agricultural phenomena. S Lardon and P-L Osty [22] have shown an application of spatial modelling on bushes expansion in farming lands. Lardon and Piveteau [23] have revised and adapted the chorem grid for territory management purposes. They distinguished structures (considered space objects) and dynamics (spatial processes in which these objects are identified). However, decision-makers have hand-extract and hand-draw their chorems from data sources.

Thus, the framework presented in the paper enhances chorems systems since it allows the online extraction and visualization of chorems using SOLAP operators. At the same time, adding chorems visualization to SOLAP improves its geovisualization analysis capabilities (c.f. Sect. 7).

4.2   Agricultural Data Analysis

Few works focus on using OLAP and SOLAP systems for the storage and analysis of agricultural data (for example [10, 29]). A survey is presented in [7]. These works mainly study farms productions (cotton, milk, etc.) highlighting multidimensional design issues such as versioning, conceptual design, etc. From an implementation point of view, OLAP-GIS integrated solutions are mostly used. The SOLAP clients support only classical geovisualization methods such as interactive thematic and cloropeth (multi) maps. However, those cartographic visualization methods and tools appear sufficient because the aggregated indicators (for example sum of cotton produced) of these spatio-multidimensional models are quite simple (limited to existing SQL aggregation functions). Indeed, as we shown in the next sections, when indicators are more complex, new geovisualization methods are needed.

The usage of chorems in agricultural context has been also few investigated. Only [21, 24, 25] study the representation of the spatial organization of agricultural activities using handmade chorems. For example the authors visually represent the proximity of plots with the farm buildings using the attraction chorem. Contrary to the agricultural context, chorems have been widely used in other domains such as land management [23], environmental analysis [26]. However, we believe that the usage of chorems in the agricultural domain rests unexplored since chorems visualizations are well adapted to agricultural data and analysis as shown in the next.

The first benefit in using chorems as geovisualization methods in agricultural data analysis is the decomposition of the concept of "yield". Indeed, the combination of production values (for example wheat tons produced) and cultivated plots (hectares cultivated) is often calculated using a single indicator (i.e. yield) leading to simplified analyses. For example, the reasons for a yield change can be found either in a better production process, or in a reduction of cultivated land caused by an important urban sprawl. In other words, the yield (tons/hectares) variation can be explained either by increasing the level of production, or by reducing the cultivated surface. Per contra, using a chorem approach is possible to combine the two parameters of yield indicator in a single visualization. This allows comparing the changes in the ratio between production and surface using one simple graphical representation for the two parameters.

The multi spatial scale visualization offered by chorem visualization also appears as an important interesting point for agricultural data analysis. Indeed, usually spatial analysis is based on different divisions of the space with agronomic consistencies: administrative limits (departments, regions, etc.), or environmental limits (e.g. forests, urban area, etc.) because these divisions have different topological characteristics (for example a forest can belong to different departments). In chorem visualization this duality is not more present since the space is represented by icon-based representations that do not use the topology of the divisions. Therefore, chorems allow data visualization at these different scales and divisions at the same time, allowing to identify similar agronomic and managing processes.

The last advantage of the usage of chorem visualization in agriculture is the potential understanding of the spatial component of data by the decision-makers, who are more sensitive to the technical and statistical dimensions than to the territorial one. Indeed, chorems natively include the most important key features of the spatial phenomena in their graphic displays. Then, all categories of decision-makers and citizens can understand and criticize the different aspects of the geographic phenomena. As consequence, chorems can be used as mediation and participatory support between and for the different stakeholders of agriculture.

5   Case Study

An example of a Spatial DW (SDW), which will be used all along the paper to describe our proposal, is depicted in Fig. 2 using the UML profile presented in [8]. Here, several stereotypes have been defined, one for each element of the SDW. For example a spatial dimension presents the <<SpatialDimension>> stereotype, a spatial level is identified with <<SpatialAggLevel>> stereotype. The <<Fact>> stereotype designs facts and numerical and spatial measures have the stereotypes <<NumericalMeasure>> and <<SpatialMeasure>>.

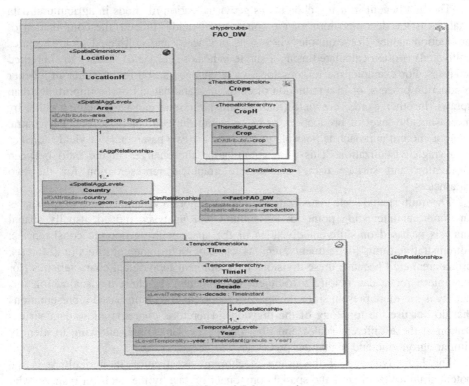

Fig. 2. FAO spatial data warehouse.

The SDW is loaded using open-data of FAO [16]. It allows analysis of agricultural cultivated surface and production per year, country and crop. It presents a spatial hierarchy grouping countries in areas, and years by decade. Using this SDW it is possible to answer queries like: "What is the total surface and production of wheat per country and year?". More complex analysis could be performed using this SDW. In particular to evaluate national agricultural policies, it is possible to compare agricultural production and surface over time, for example using the query: "What are differences of total surface and production per country on the last 5 years. In our case study, we work using national wheat production and national wheat area harvested of European countries between 1991 and 2011. These data allow us to analysing not only the variations of production and acreage but also variation of crop yields and productivity. The range of the period studied lets us to do several analysis in different temporal scales.

6 ChoremOLAP

In this section, we present the theoretical framework for integration of SOLAP and chorems (Sect. 6.2), and its implementation in a SOLAP tool (Sect. 6.1) (https://www.youtube.com/watch?v=srx0Hm7BPkw).

6.1 Architecture

ChoremOLAP architecture is described in Fig. 3. It is based on a Relational SOLAP architecture composed of three tiers: SDW, SOLAP server and SOLAP client.

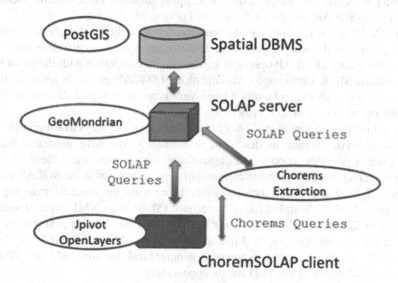

Fig. 3. ChoremOLAP architecture.

The Spatial Data Warehouse tier is implemented using the Spatial DBMS PostGIS [34]. PostGIS is an extension of Postgres providing a native support for spatial data and spatial analysis functions. This tier is used for storing alphanumeric and spatial multidimensional data. Warehoused spatial data is stored using the star-schema [31], where levels of the denormalized spatial dimension present geometric attributes.

Example. The logical model of our case study is shown on Fig. 4, where a table for each dimension is defined, and one table for the fact.

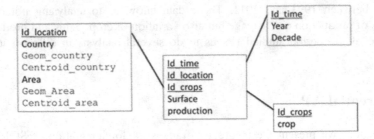

Fig. 4. Star-schema of the FAO SDW.

The spatial dimension presents a geometry column for each spatial level (Geom_country and Geom_area). We also note that two additional geometries representing the centroid of the countries and the areas have been defined since, as detailed in the next sections, they are used for the chorems visualization.   □

The SOLAP server used is GeoMondrian [18]. GeoMondrian is an open-source SOLAP Server supporting GeoMDX. GeoMondrian represents dimensions and measures using an XML file, which defines a mapping on the logical schema. GeoMondrian supports SOLAP queries on the top of Postgis.

The SOLAP client is a web-based client composed of the OLAP client JPivot and the GIS client OpenLayers. In particular, JPivot is an open-source web-based OLAP client implementing all OLAP operators by the simple interaction with the pivot table. JPivot supports MDX. Cartographic visualization of SOLAP queries is provided by the cartographic web client OpenLayers. Openlayers is an open source JavaScript library for displaying map data in web browsers.

The architecture presented in [5] has been extended in two ways to support chorems extraction and visualization as described in Sect. 6.3. We have used this SOLAP system since it provides an open and customizable visual interface. Indeed, the main idea for implementing customizable cartographic visual displays in the SOLAP client is the usage of SLD and GML standards, which are used by standard mapping web services (e.g., WMS). A Styled Layer Descriptor (SLD) is an XML schema specified by the Open Geospatial Consortium (OGC) for describing the appearance of map layers. Moreover, the Geography Markup Language (GML), defined by the Open Geospatial Consortium, allows expressing geographical features. We use GML to represent spatial data and the SLD for its appearance.

The original geovisualization proposed method in [5] consists of chloropleth maps (e.g. coloured geometries) implemented using GML and SLD. Here, we have added the visualization of icons representing chorems as described in the Sect. 6.2.

In the SOLAP server, we have implemented a component that translates a chorem query in a classical SOLAP GeoMDX query. In this way, chorem queries are transparently handled by any SOLAP server. The proposed extensions are detailed in the rest of this section.

6.2   Principles

Our geovisualization methods are based on two main groups:

Chorem-based geovisualization methods, which are based on chorems, and *Non chorem-based geovisualization methods*, which are classical geovisualization methods.

6.2.1   Chorem-Based Geovisualization Methods Principles

The chorems used in our approach are a subset of chorems identified by [22] that can be extracted from spatial warehoused data, as shown on Fig. 5. Here, 5 main groups of chorems are described.

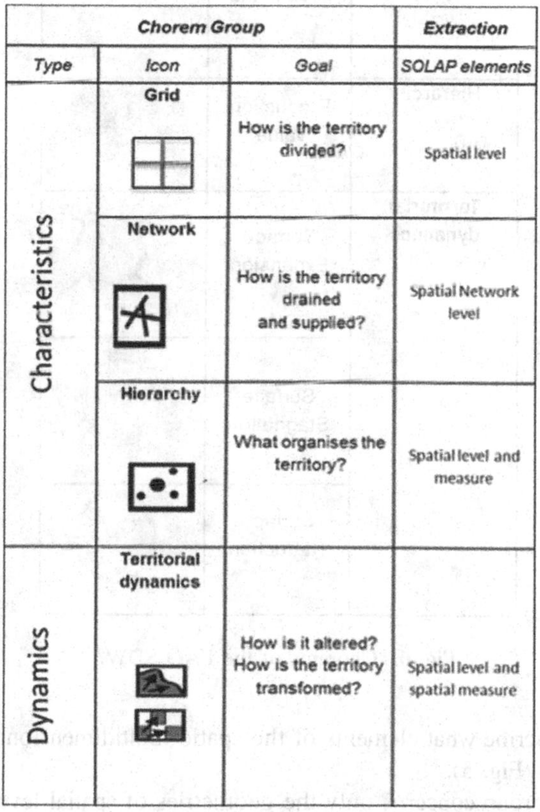

Chorem Group			Extraction
Type	Icon	Goal	SOLAP elements
Characteristics	Grid	How is the territory divided?	Spatial level
	Network	How is the territory drained and supplied?	Spatial Network level
	Hierarchy	What organises the territory?	Spatial level and measure
Dynamics	Territorial dynamics	How is it altered? How is the territory transformed?	Spatial level and spatial measure

Fig. 5. Chorems and extraction mapping of the ChoremOLAP framework.

In particular, Grid represents how the territory is divided by actors (for example, municipalities). Network designs the presence of network structures such as roads, rivers, but also informational networks drained and supplied the territory. Hierarchy specifies the different entities and how they organize the territory.

The dynamic chorems result from the temporal evolution of these structures. The Territorial Dynamic transforms differentiated spaces, even by continuous expansion or by discrete allocation.

Example. In Fig. 6 an example of instance for each group of chorems is also presented using our case study.   □

Group	Instance	
	Name	Icon
Grid	Area limits	
Hierarchy	Production value	
Territorial dynamics	Surface Expansion	
	Surface Stagnation	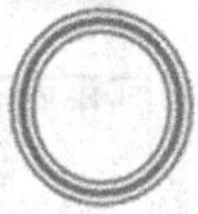
	Surface Reduction	

Fig. 6. Chorems of the FAO SDW.

Let us now describe what elements of the spatio-multidimensional model are used to extract chorems (Fig. 5).

Grid chorems group concerns only the geometries of spatial levels.

Example. In our case study the chorem "Area limits" is simply defined using the spatial level "Area" of the SDW.   □

In the same way, Network chorems group is also only associated to spatial levels defined as spatial network levels in [6].

Hierarchy chorems group refers to spatial members and their numerical properties, which can change along non-spatial dimensions.

Example. In our example, "Production increase" chorem is defined using the spatial level "Country" and the measure "Production".   ☐

Finally, Territorial Dynamics chorems group is similar to Hierarchy, but here the properties are strictly related to the geometries of the spatial level (spatial measures).

Example. The "surface" measure is a spatial measure, and so the "Surface Reduction" chorem is calculated using the spatial level "City" and "surface" measure.   ☐

6.2.2 Non Chorem-Based Geovisualization Methods Principles

The system proposed in [5] allows to displays results of SOLAP queries using choropleth maps. Here we extended them by using the simplified geometries of the Grid chorems group.

Moreover, it allows visualizing nominal measures using an iconic representation. For example, in order to visualize production evolution, we define three icons: ● if there is an augmentation, ● if there is a diminution, and ● otherwise.

6.3 Extraction and Visualization

In this section, we detail how chorems of Fig. 5 are extracted (Sect. 6.3.1) and visualized (Sect. 6.3.2) on the top of the SOLAP architecture of Fig. 3

6.3.1 Extraction

In order to extract chorems on the top of a classical SOLAP server, we use MDX. MDX is de-facto standard query language of OLAP servers. MDX allows defining calculated measures (i.e. measures calculated using measures values stored in the SDW tier).

The template MDX formula for the Territorial Dynamics chorem is presented in Fig. 7. The chorem is represented by the calculated measure [Measures]. [ChoremE] that assumes the values:

"−1" the phenomenon reduces
"0": the phenomenon does not change
"+1" the phenomenon expands

where:

Phenomena is the measure used for the chorem definition. For example Phenomena = "Surface" allows the extraction of Surface Reduction, Surface Stagnation and Surface Expansion chorems respectively (Fig. 7);

TimeRange represents the interval between two dates (Date) (for example TimeRange = 5 allows comparing Phenomena values of each year with 5 years ago values).

6.3.2 Visualization

The visualization of the Grid chorems group is simply achieved by the visualization of simplified geometries of the spatial levels members stored in the SDW tier (Fig. 5).

Example. An example of Grid visualization is shown on Fig. 8.   ☐

```
with member [Measures].[Y-1] as
'IIf(((ParallelPeriod([Time].[year], TimeRange,
[Time].CurrentMember) IS NULL) OR
(ParallelPeriod([Time].[Date], 1.0, [Time].CurrentMember) =
0.0)), (- 1.0), ParallelPeriod([Time].[Date], 1.0,
[Time].CurrentMember))'
```
//check for existing previous dates

```
  member [Measures].[diff] as 'IIf(([Measures].[Y-1] > (-
1.0)), ([Measures].[Phenomena] - [Measures].[Y-1]), (-
1.0))'
```
//calculate the difference with previous date

```
  member [Measures].[TauxVariation] as 'IIf(([Measures].[Y-
1] > (- 1.0)), ([Measures].[diff] / [Measures].[Y-1]), (-
1.0))', format_string = "Percent"
```
//transform the difference in a percentage value

```
  member [Measures].[Choreme] as 'IIf(([Measures].[Y-1] <
0.0), "-2", IIf(([Measures].[TauxVariation] < 0.0), "-1",
IIf(([Measures].[TauxVariation] > 0.0), "+1", "0")))'
```
//calculate the chorem values

Fig. 7. Territorial Dynamics chorem MDX template.

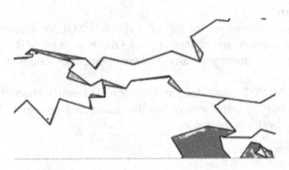

Fig. 8. Visualization of the Grid chorem.

The visualization of the Territorial Dynamics chorems group is implemented using a simple SLD template (Fig. 9).

An SLD template is generated for each combination of non spatial members (NonSpatialDimensionsMembersChorem) present in the pivot table result (for example "2000"). It also defines a rule for each chorem value ChoremValue corresponding to the calculated measure [Measures].[ChoremE] (−1, 0, +1). For each rule, an image associate to the chorem value is visualized (ChoremImage). The location where this image is displayed is represented using GML. We use the centroid of the spatial member, when it is a polygon stored in the spatial dimension table (Fig. 4).

```
<Rule>
    <ogc:Filter>
        <ogc:PropertyIsEqualTo>
            <ogc:PropertyName>NonSpatialDimensionsMembersChorem</ogc:Property
        Name>
            <ogc:Literal>ChoremValue</ogc:Literal>
        </ogc:PropertyIsEqualTo>
    </ogc:Filter>
    <PointSymbolizer>
        <Graphic>
            <ExternalGraphic>
                <OnlineResource xlink:href="ChoremImage" xlink:type="simple"/>
                <Format>image/gif</Format>
            </ExternalGraphic>
            <Size>ChoremImageSize</Size>
        </Graphic>
    </PointSymbolizer>
    <PolygonSymbolizer>
        <Fill>
        <CssParameter name="fill-opacity">0.0</CssParameter>
        </Fill>
    </PolygonSymbolizer>
</Rule>
```

Fig. 9. Territorial Dynamics chorem SLD template.

Fig. 10. Surface stagnation, surface reduction, and surface expasion.

Example. Figure 10 shows an example of the chorems Surface Stagnation, Surface Reduction, and Surface Expansion for each country in 2000. Decision-makers rapidly "see" that Italy presents a surface reduction, while France has a surface expansion. □

The Hierarchy chorems group is implemented in the same way, but here we present only an example of visualization.

Example. An example of Hierarchy chorem visualization of the production on 2000 is shown on Fig. 11 (Italy and Switzerland are in the same category, while France production is higher). □

An example of non-chorem based geovisualization (Sect. 6.2.2) showing a chloropleth map for the value of the surface on 2000 and the production evolution per country is shown on Fig. 12.

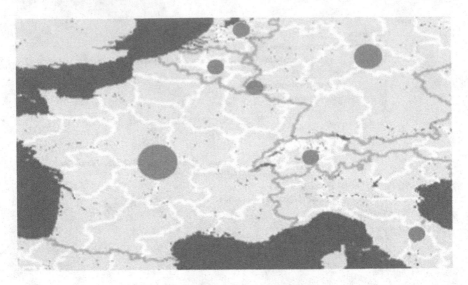

Fig. 11. Production value.

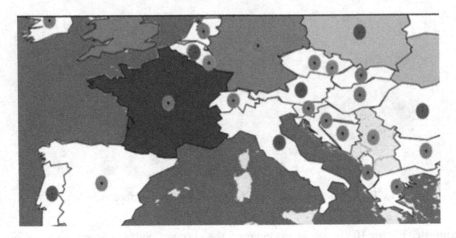

Fig. 12. Non chorem-based geovisualization.

7 Discussion

In this section, we analyse how our approach mutually improves SOLAP and chorems systems.

7.1 Chorems Improved by SOLAP

As we state in the previous section, ChoremOLAP allows to interactively creating chorems. This is achieved by simply triggering SOLAP queries, as described in the following.

SOLAP operators allow to explore the warehoused data on-line aggregating measures values, and in our tool also chorem values. For example, decision-makers can move from the "Area" spatial level to the "Country" spatial level by the simply interaction with the pivot table of our web-client (i.e. DrillDown operation on the spatial dimension) (Fig. 13). As shown on Fig. 13(a), the surface chorem is visualized at the "Area" spatial level. When the decision-maker DrillDowns to the "Country" spatial level, the chorem map is instantaneously re-calculated for each country (Fig. 13(b)).

	Time
	2000
	Measures
Location	Area_Harvested
Eastern Europe	35274681
Northern Europe	4008100
Southern Europe	7026760
Western Europe	8967529

- ☑ind_2000_Area_Harvested - 4008100 ☐ 10261416 ▨ 16514732 ▨ 22768048 ▨ 29021364 ▨ 35274681
- ☑ind_2000_AreaChorem1 - ⌥ Expansion - ○ Stagnation - ⊖ reduction

(a) Drill Down for automatic chorem extraction and visualization

	Time
	2000
	Measures
Location	Area_Harvested
Belarus	452000
Bulgaria	978575
Czech Republic	970435
Czechoslovakia	
Hungary	1024430
Poland	2635097
Republic of Moldova	372967
Romania	1928328
Russian Federation	21346000
Slovakia	405249
Ukraine	5161600

- ☑ind_2000_Area_Harvested - 372967 ☐ 4567573 ▨ 8762179 ▨ 12956785 ▨ 17151391 ▨ 21346000
- ☑ind_2000_AreaChorem1 - ⌥ Expansion - ○ Stagnation - ⊖ reduction

(b) Drill Down for automatic chorem extraction and visualization (a).

Fig. 13. (a) Drill down for automatic chorem extraction and visualization (b) Drill down for automatic chorem extraction and visualization

In the same way, the decision-maker can dynamically change other dimensions. For example starting from the chorem map of Fig. 13(a), he change the year, for example moving from 2000 to 1995, and the chorem map is online calculated.

Thus, we can conclude that SOLAP system allows the online creation and visualization of chorem maps.

7.2   SOLAP Improved by Chorems

Let us now describe how SOLAP maps are improved by chorems visualization. To evaluate the new analysis capabilities offered by our framework from a visualization point of view, we performed a comparative study of ChoremOLAP against one of the most advanced commercial SOLAP clients. We compare our proposal to classical SOLAP visualization methods and we analyze the ability to represent different kind of SOLAP queries.

In Table 1 we present what and how many measures can be visualized with geovisualization methods for SOLAP including our chorem Fig. 13 maps.

Table 1. Geovisualization methods for SOLAP.

SOLAP Geovisualization methods		Measure type	
		Number	Type
Thematic map (map using diagram displays such as bar charts, pie charts)		More than 1	Quantitive
Cloropeth map (map using only one colour per geographic objects)		1	Ordinal
			Nominal
Cloropeth Multimaps (set of cloropeth maps displayed in the same screen)		More than 1	Ordinal
			Nominal (one per map)
Thematic Multimaps (set of thematic maps displayed in the same screen)		More than 1	Quantitive
Chorem map	Hierarchy	1	Ordinal
	Territorial dynamics	1	Nominal

In Table 2 we evaluate these geovisualization methods on 6 SOLAP queries. These queries represent all possible combinations of possible measures involved in a SOLAP query.

Query Q1 is a classical SOLAP query, therefore there is no need to use chorems.

Table 2. Evaluation of geovisualization methods.

Query	Measures	Measures type	SOLAP geovisualization	Our Framework
Q1 "What is the total surface and production of wheat per country and year?"	two classical SOLAP measures	2 quantitative	Thematic map	
Q2: "What is difference of total production per country on the last 5 years on 2000?"	One chorem (production evolution)	ordinal	Cloropeth map	Hierarchy chorem map
Q3: "What is difference of total surface per country on the last 5 years on 2000?"	One chorem (surface evolution)	nominal	Cloropeth map	Territorial dynamics chorem map
Q4: "What is difference of total surface per country on the last 5 years on 2000, and the production on 2000?"	One chorem (surface evolution)	nominal	Cloropeth map	Territorial dynamics chorem map
	Classical SOLAP measure	quantitative	Thematic map	Thematic map
Q5: "What is difference of total production per country on the last 5 years on 2000, and the production on 2000?"	One chorem (production evolution)	ordinal	Cloropeth map	Hierarchy chorem map
	Classical SOLAP measure	quantitative	Thematic map	Thematic map
Q6: "What are differences of total surface and production per country on the last 5 years on 2000?"	One chorem (surface evolution)	nominal	Cloropeth Multimaps	Hierarchy chorem map
	One chorem (production evolution)	ordinal		Territorial dynamics chorem map

Query Q3 concerns one chorem (surface evolution). Here using chorem visualization is very satisfying for the decision-maker since to each nominal value of the chorem (stagnation, etc.) a particular icon is used. Chloropleth map can be also used, but decision-maker is forced to mentally associate a colour to a surface evolution.

Thus, chorem maps should be preferable to chloropleth maps for nominal measures.

For the query Q2 chorem maps and chloropleth maps have the same expression power.

For the Queries Q4 and Q5, our framework presents the same advantage of Query 3.

Query Q6 concerns 2 chorems (surface and production evolution). Therefore, chloropleth multimaps (one chloropleth map per measure) can be used, but the main limitation is that the decision-maker has mentally to overlay the maps to compare the two measures country by country. Thus our approach seems perform chloropleth multimaps.

To conclude, our geovisualization methods based on chorems do not always replace classical geovisualization methods of SOLAP tools, but they appear useful when dealing with phenomena that can be represented as chorems.

However, usability test should be provided to quantify the advantage of using chorem maps instead of SOLAP maps. They represent our future work.

8 Conclusions and Future Work

Spatial OLAP systems are decision-support systems allowing analysing huge volume of geo-referenced data. SOLAP systems provide clients with interactive graphic, tabular and cartographic displays. In particular, geovisualization methods used by existing SOLAP systems are limited to interactive cloropeth and thematic maps. However, when the analysed spatial phenomena are complex, advanced geovisualization techniques are need. Recently some geovisualization methods and tools have been developed using chorems. Theoretically, a chorem is a visual summary of a geographic phenomenon. However, chorem systems are based on pre-defined maps, which limit potentiality of spatial decision-making process.

Then in this paper we present a framework and its implementation in a SOLAP architecture for augmenting the analysis capabilities of the SOLAP clients with chorem maps.

In detail, we propose a set of methods to on-line extract and visualize chorems on the top of a SDW. We also propose an implementation of our framework using a general architecture based on standards.

Our proposal is presented using a case study concerning the analysis of agricultural data. Therefore, all along the paper we describe what are issues related to the analysis of agricultural data, and how our system meets those requirements.

As future work, we plan to investigate other chorems as defined in [23]. We also plan to define a usability study to evaluate in a quantitative way the pro and cons of the usage of chorems instead of classical SOLAP geovisualization methods from a visualization point of view.

References

1. Bédard, Y., Rivest, S., Proulx, M.-J.: Spatial on-line analytical processing (SOLAP): concepts, architectures, and solutions from a geomatics engineering perspective. In: Data Warehouses and OLAP: Concepts, Architecture, and Solutions, vol. 14, pp. 298–319 (2006)
2. Bédard, Y., Proulx, M., Rivest, S., et al.: Merging hypermedia GIS with spatial on-line analytical processing: towards hypermedia SOLAP. In: Stefanakis, E., Peterson, M.P., Armenakis, C., Delis, V. (eds.) Geographic Hypermedia: Concepts and Systems. Springer, Berlin (2006)
3. Bimonte, S., Tchounikine, A., Di Martino, S., Ferrucci, F.: Supporting geographical measures through a new visualization metaphor in spatial OLAP. ICEIS (5), 19–26 (2007)
4. Bimonte, S.: A web system for spatio-multidimensional analysis of geographic and complex data. Int. J. Agric. Environ. Inf. Syst. 1(2), 42–67 (2010). IGI Global

5. Bimonte, S.: A generic geovisualization model for spatial OLAP and its implementation in a standards-based architecture. Ingénierie des Systèmes d'Information **19**(5), 97–118 (2014)

6. Bimonte, S., Kang, M.-A., Trujillo, J.: Integration of spatial networks in data warehouses: a UML profile. In: Murgante, B., Misra, S., Carlini, M., Torre, C.M., Nguyen, H.-Q., Taniar, D., Apduhan, B.O., Gervasi, O. (eds.) ICCSA 2013, Part IV. LNCS, vol. 7974, pp. 253–267. Springer, Heidelberg (2013)

7. Bimonte, S.: Spatial OLAP for agri-environmental data and analysis: lessons learned. In: MIPRO 2015, pp. 1393–1398 (2015)

8. Boulil, K., Bimonte, S., Pinet, F.: Conceptual model for spatial data cubes: a UML profile and its automatic implementation. J. Comput. Stand. Interfaces **38**, 113–132 (2015)

9. Brunet, R.: La carte-modele et les choremes. Mappemonde **4**, 4–6 (1986)

10. Chaturvedi, K., Rai, A., Dubey, V., Malhotra, P.: On-line analytical processing in agriculture using multidimensional cubes. Int. Indian Soc. Agric. Stat. J. **62**(2), 56–64 (2008)

11. De Chiara, D., Del Fatto, V., Laurini, R., Sebillo, M., Vitiello, G.: A chorem-based approach for visually analyzing spatial data. J. Vis. Lang. Comput. **22**(3), 173–193 (2011)

12. Deffontaines, J.P., Cheylan, J.P., Lardon, S.: Gestion de l'espace, des pratiques aux modèles. Mappemonde **1990**(4) (1990)

13. Del Fatto, V.: Visual summaries of geographic databases by Chorems. Ph.D. thesis (2009)

14. Del Fatto, V., Laurini, R., Lopez, K., Sebillo, M., Vitiello, G.: A chorem-based approach for visually synthesizing complex phenomena. Inf. Vis. **7**(3–4), 253–264 (2008)

15. Di Martino, S., Bimonte, S., Bertolotto, M., Ferrucci, F.: Integrating google earth within OLAP tools for multidimensional exploration and analysis of spatial data. In: Filipe, J., Cordeiro, J. (eds.) ICEIS 2009. LNBIP, vol. 24, pp. 940–951. Springer, Heidelberg (2009)

16. FAO (2015). http://data.fao.org/statistics

17. Golfarelli, M., Mantovani, M., Ravaldi, F., Rizzi, S.: Lily: a geo-enhanced library for location intelligence. In: Bellatreche, L., Mohania, M.K. (eds.) DaWaK 2013. LNCS, vol. 8057, pp. 72–83. Springer, Heidelberg (2013)

18. Geomondrian (2015). http://www.spatialytics.org/fr/projets/geomondrian/

19. Inmon, W.: Building the Data Warehouse. QED Information Sciences Inc., Wellesley (1992)

20. Intelli3. http://www.intelli3.com/en/map4decision_en

21. Lardon, S., Capitaine, M., Benoît, M.: Les modèles graphiques pour représenter l'organisation spatiale des activités agricoles. In: Représentations graphiques dans les systèmes complexes naturels et artificiels. 9° Journées de Rochebrune: Rencontres Interdisciplinaires sur les Systèmes Complexes Naturels et Artificiels. ENST Editions, Paris, pp. 127–150 (2000)

22. Lardon, S., Osty, P. L.: Time-space dimensions of farmers' practice: methodological proposals from surveys and modelling of sheep farming. In: Case Studies in Southern Massif Central, France, Fourth European Symposium. European Farming and Rural Systems Research and Extension into the Next Millenium. Environmental, Agricultural and Socio-economic Issues (2000)

23. Lardon, S., Piveteau, V.: Méthodologie de diagnostic pour le projet de territoire: une approche par les modèles spatiaux. Géocarrefour: Revue de Géographie de Lyon **80**(2), 75–90 (2005)

24. Lardon, S., Le Ber, F., Metzger, J.-L., Osty, P.-L.: Une démarche et un outil pour modéliser des organisations spatiales agricoles et raisonner à partir de cas d'exploitations agricoles. Revue Internationale de Géomatique **15**(3), 263–280 (2005)

25. Lardon, S.: La modélisation graphique. In: Benoît, M., Deffontaines, J.P., Lardon, S. (eds.) Acteurs et territoires locaux. Vers une géoagronomie de l'aménagement. Editions INRA, Savoir faire, pp. 33–55 (2006)

26. Lebourg, M.-N., Lardon, S., Cot, C., De Nayer, A., Lacroix, C., Dumont, A.: La méthode de diagnostic partagé: comprendre et analyser un territoire avec les représentations spatiales schématiques pour produire le « Dire de l'État », AgroParisTech Clermont-Ferrand et DREAL Poitou-Charentes (2014)

27. MacEachren, A., Gahegan, M., Pike, W.: Geovisualization for knowledge construction and decision support. IEEE Comput. Graph. Appl. **24**(1), 13–17 (2004)

28. Malinowski, E.: GeoBI architecture based on free software. In: Geographical Information Systems Trends and Technologies. Elaheh Pourabbas CRC Press (2014)

29. Nilakanta, S., Scheibe, K., Rai, A.: Dimensional issues in agricultural data warehouse designs. J. Comput. Electron. Agric. **60**(2), 263–278 (2008)

30. Keim, D.A., Mansmann, F., Schneidewind, J., Ziegler, H.: Challenges in visual data analysis. In: Information Visualization (IV 2006). IEEE Press (2006)

31. Kimball, R.: The Data Warehouse Toolkit: Practical Techniques for Building Dimensional Data Warehouses. Wiley, Hoboken (1996)

32. Klippel, A., Tappeb, H., Kulikc, L., Leed, P.: Wayfinding choremes a language for modeling conceptual route knowledge. J. Vis. Lang. Comput. **16**(4), 311–329 (2005)

33. Pestana, G., da Silva, M., Bédard, Y.: Spatial OLAP modeling: an overview base on spatial objects changing over time. In: IEEE 3rd International Conference on Computational Cybernetics, 13–16 avril, Mauritus (2005)

34. PostGIS (2015). http://postgis.net/

35. Shekhar, S., Gunturi, V., Evans, M.R., Yang, K.: Spatial big-data challenges intersecting mobility and cloud computing. In: MobiDE 2012, pp. 1–6 (2012)

36. Silva, R., Moura-Pires, J., Santos, M.: Spatial clustering in SOLAP systems to enhance map visualization. Int. J. Data Warehouse. Min. **8**(2), 23–43 (2012)

37. Spinsanti, L., Ostermann, F.: Automated geographic context analysis for volunteered information. Appl. Geogr. **43**, 36–44 (2013)

Training Model Trees on Data Streams
with Missing Values

Olivier Parisot[(✉)], Yoanne Didry, Thomas Tamisier, and Benoît Otjacques

Luxembourg Institute of Science and Technology (LIST), Belvaux, Luxembourg
olivier.parisot@list.lu
http://www.list.lu

Abstract. Model trees combine the interpretability of decision trees with the efficiency of multiple linear regressions making them useful in dynamically attaining predictive analysis on data streams. However, missing values within the data streams is an issue during the training phase of a model tree. In this article, we compare different approaches to deal with incomplete streams in order to measure their impact on the resulting model tree in terms of accuracy. Moreover, we propose an online method to estimate and adjust the missing values during the stream processing. To show the results, a prototype has been developed and tested on several benchmarks.

Keywords: Data streams · Model trees · Missing values imputation

1 Introduction

Model trees are convenient to predict numerical values from past observations [28,40]. In fact, they are close to decision trees: they use a formalism that is intuitive and understandable by domain experts [23]. Even if the accuracy is still the most critical aspect, the interpretability is important in many applications of predictive modeling [31].

A model tree can be defined as follows: given a data stream which is an ordered sequence of observations and where each observation is defined by n features $F_1, \ldots, F_n$, a model tree aims at evaluating the value of a continuous feature F_i ($1 \leq i \leq n$) according to the values of the other features F_j ($j \neq i$, $1 \leq j \leq n$). In terms of structure, a model tree is a directed graph composed of nodes, branches and leaves (Fig. 1). Each node is followed by branches that specify a test on the feature value (for instance: $F_1 = value$), and each leaf corresponds to a multiple linear regression model that aims at computing the continuous value to predict (Table 1).

In practice, model trees can be obtained from *static data sets* by applying the classical M5 algorithm [28], and new techniques are frequently proposed [12]. To train a model tree from a stream, an online algorithm has been proposed [16]: as a consequence, it is possible to dynamically update predictive models from data sources that are continuously refreshed like networks traffic information, financial quotes or sensors data.

© Springer International Publishing Switzerland 2016
M. Helfert et al. (Eds.): DATA 2015, CCIS 584, pp. 81–97, 2016.
DOI: 10.1007/978-3-319-30162-4_6

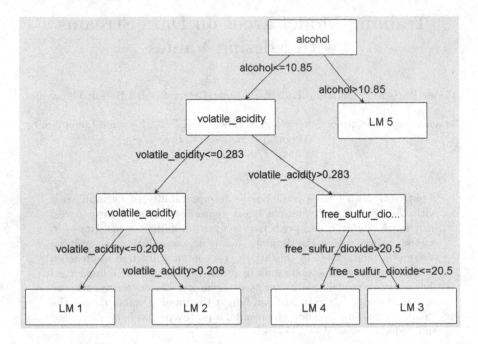

Fig. 1. A simplified version of the model tree that predicts the quality of Portuguese Vinho Verde white wines by using physico-chemical data [5]: each leaf corresponds to a multiple linear regression model that evaluates the wine quality (Table 1)

Once built, model trees are characterized as follows:

(*a*) The complexity of a model tree can be measured by its size (i.e. the number of nodes) [4]. This indicator is important because it may indicate *over-learning* [32]. In addition, a large model tree is hard to visualize and interpret [34].

(*b*) The accuracy of a model tree is its ability to predict correct values, i.e. the difference between predicted and expected values. Traditionally, it can be estimated by considering the data as two parts (training set and evaluation set). In the context of data streams, the accuracy has to be measured iteratively for each observation of the considered stream. To this end, various metrics exist like the Mean Absolute Error (MAE) and the Root Mean Squared Error (RMSE).

A simple example of predictive model tree design is the following: given a dataset that describes the physico-chemical properties of a set of Portuguese Vinho Verde red wines, it is possible to predict the quality of these wines [1,5]. The wine quality is defined by a numerical value that can be considered as a score: 0 represents a poor wine and 10 represents an excellent wine. To this end, a model tree can be built from these data in order to *predict* the quality of new wines by using the physico-chemical data. From a graphical point of view, model trees are often represented using node-link diagrams, even if other

Table 1. Multiple linear regression model for each leaf of the model tree that predicts the quality of Portuguese Vinho Verde white wines (Fig. 1). The estimated value is a numerical score between 0 and 10 (0 for a poor wine, 10 for an excellent wine) [5].

Leaf of the model tree	Regression model to evaluate the wine quality
LM1	− 0.1122 * volatile_acidity + 0 * free_sulfur_dioxide + 0.0049 * alcohol + 6.0006
LM2	− 0.0788 * volatile_acidity + 0 * free_sulfur_dioxide + 0.2372 * alcohol + 3.3594
LM3	− 0.0442 * volatile_acidity + 0.0003 * free_sulfur_dioxide + 0.0037 * alcohol + 5.047
LM4	− 0.0442 * volatile_acidity + 0.0001 * free_sulfur_dioxide + 0.0037 * alcohol + 5.3184
LM5	− 0.0156 * volatile_acidity + 0.0121 * free_sulfur_dioxide + 0.3269 * alcohol + 2.0913

representations like concentric circles are possible [27]. In this example, using this kind of representation helps to highlight that four properties can be used to predict the wine quality: *alcohol*, *sulphates*, *fixed acidity* and *volatile acidity* (Fig. 1, Table 1).

Unfortunately, data quality is often a crucial issue in many real world applications, especially when data are incomplete [11,44]. As an example, during the training of a model tree, the learning algorithms can not deal with incomplete data directly, so the data stream has to be preprocessed as a prior step. Ideally, the imputation should be realized and controlled in order to have a positive benefit for further usage [9].

To tackle this issue, we compare in this paper different approaches to deal with incomplete streams, and we propose an online method to adjust the missing values estimation in such a way that it tends to increase the trained model tree accuracy.

The rest of this article is organized as follows. Firstly, related works about model trees and missing values are mentioned. Then, a method is described in details in order to solve the identified issues. Finally, the developed software is presented and the results of experiments are discussed.

2 Related Works

2.1 Alternative to Model Trees

A multitude of techniques exists for predictive analytics, and the main issue for the *data scientist* is to select the appropriate technique according to the situation [39]. The choice mainly depends of the nature of the data (for instance: linear or non-linear relationships). From a general point of view, model trees have several advantages over the other techniques:

- Model trees are not considered as *black box* models like Artificial Neural Networks (some works have been done to explain them [10], but they still remain very complex to interpret).
- Model trees implicitly performs features selection, in order to find the characteristics that help to build the prediction. In fact, the top nodes on which the tree is split are essentially the most important variables within the *class* attribute of the dataset [3].
- Structure of model trees is easy to visualize and interpret, like for decision trees and regression trees [20]: the main rules are explicitly described by the branches of the trees, and the possible predicted values are obtained by using the formula of the leaves.

Depending of the final use-case, various alternatives exist for building predictive models [22]. To estimate continuous values, model trees are efficient and they are often easier to interpret than the other types of models.

2.2   Model Trees and Data Streams

Model trees can be trained from data streams by using the FIMT algorithm (*Fast Increment Model Trees* [16]) or the more recent FIMT-DD algorithm (*Fast Increment Model Trees with Drift Detection*) [17].

These algorithms are inspired by VFDT (*Very Fast Decision Tree*), a decision tree induction method for data streams: VDT aims at guarantying that the produced tree will be asymptotically close to the batch tree given that enough examples are analyzed [7]. To do that, VFDT is based on the Hoeffding's probability inequalities [15], and they can be applied successfully on large datasets too [14].

Nevertheless, these techniques do not manage the case of incomplete data streams and there is a need to deal with missing values [11, 26, 33, 45].

2.3   Dealing with Missing Values

Dealing with missing value is a well-known topic in data mining, especially when predictive models have to be built on incomplete data. Depending of the nature of the data absence (*missing completely at random, missing at random, missing not at random*) [29], several strategies can be applied:

- Observations with missing data can be simply deleted/ignored: this trivial approach can be sufficient in a lot of cases.
- Missing values can be estimated and these estimations can be used to train the predictive models.
- Robust learning methods can be applied in order to minimize the impact of missing data.

In the next paragraphs, we analyze the pros and cons of these approaches.

Ignoring Observations with Missing Values. The first naive approach simply discards the observations with missing values [8]. This solution is straightforward to apply but has a major drawback: if a lot of data are missing from the stream - which is quite frequent in real-world cases [11], then the model tree will be trained with few observations. In addition, it creates a bias in predictive models if the values are systematically missing in certain situations (as *missing not at random*) [29].

Estimating Missing Values. In order to resolve this issue, data preprocessing clearly helps to improve the performance of learning algorithms [9,43]. In the literature, various methods have been proposed to estimate missing values [21,37]:

- A simple solution consists in replacing missing numerical values by the mean values, and is still used in many statistical software packages. However, this can highly disrupt the data structure and so degrade the performance of the statistical modeling [19].
- Regression methods can be used for this task, especially when obvious relationships between the attributes are known. In addition, regression trees are good candidates because they are efficient and easy to interpret like decision trees [20].
- Artificial neural networks like multilayer perceptrons [35] or Self Organized Maps [24] have been recently used to preprocess missing hydrological data. Even though some works have been proposed to render them more interpretable [10], they still tend to have a bad reputation, as they are often considered as *black box* models. Nevertheless, they represent an extremely helpful approach building powerful predictive models, capable of providing satisfactory results.
- The Expectation-Maximization algorithm can be used to impute missing data with a good confidence level by estimating missing values with a multivariate normal model [38].
- Nearest-neighbors technique can be applied as follows: for each incomplete record, similar records are identified (by using a distance formula for a *brute force* search: Euclidean, Manhattan, Mahalanobis, etc.) and then used to estimate missing values.

However, processing data streams requires to apply *time efficient* solutions, without accessing to historical data. As a result, the classical imputation techniques can be applied on streams by using a certain pool of observations (by using a sliding window, for example) [44] or by applying specific online/incremental methods.

Recently, an incremental method has been proposed to cleverly replace missing values depending of the data type and the data distribution [26]. An other one suggests to use *ensemble learning* to fix this issue in data [42].

In our case, training a model tree on a incomplete data stream can be done as follows: it consists in filling incomplete observations with estimated values during the training of the model tree. As discussed previously, a simple application of this approach can be done by replacing missing value with the means. The imputation method can be implemented too with a set of methods dedicated to data streams, for example: (*a*) decision trees for imputing categorical missing values [7], (*b*) regression trees or model trees for imputing numerical missing values [16,18].

Unfortunately, these approaches do not take into account the impact of using these corrected data for the training phase: in other words, they may affect positively or negatively the model tree accuracy. To the best of our knowledge, this aspect is rarely studied in the literature, and we only found few papers that discussed about the consequences of the missing data estimation on the learned predictive models [30,33].

Robust Models to Deal with Incomplete Data Streams. It is not always possible or desirable to impute missing values (for example, if the values are systematically missing or if it is technically impossible to realize estimations). In these cases, it is preferable to apply robust models on the incomplete data streams, i.e. techniques that tolerates *uncertain* streams [36]. To this end, a lot of techniques are described in the literature: for example, a recent method was proposed to build robust regression models for streams [45].

Very often, these robust techniques are based on *bagging* (or *bootstrap aggregating*): originally, they aim at reducing the predictive models variance by splitting data into different samples and defining a model for each of them. In practice, they help to manage data missingness by cleverly building samples without missing values [30].

For this paper, we did not consider these *bagging* techniques and we only focused on iterative methods that handle missing values by wholly processing the data streams.

3   Contribution

During the training of a model tree from a data stream, we have seen that it is mandatory to apply a strategy to deal with missing values.

In the same time, it's critical to control the impact of the applied strategy on the learned model [9]. If we want to build a predictive model based on model trees, then the question is the following: what are the effects of using estimated values to train the model tree? In fact, it can impact positively or negatively the model tree size (i.e. the interpretability) and the model tree accuracy (i.e. the prediction error).

The inputs of the problem are the following: (a) A data stream where each observation is defined by n features $F_1, \ldots, F_n$. (b) A continuous class feature F_c to predict with a model tree: F_c. (c) A feature with possibly missing values: F_m ($m \neq c$). We assume that this feature is *important* for the prediction, i.e. it is likely used in the leading model tree to predict the value F_c (this hypothesis was already described in a work about the computation of classification models for incomplete static datasets [30]).

Given these inputs, various strategies can be applied to train a model tree to predict F_c and we have selected these ones:

(a) Skipping observations for which data are missing for F_m (Algorithm 1).
(b) Simple estimation of missing data by using the mean value for F_m (computed in an online way) (Algorithm 2).
(c) Advanced estimation of missing data for F_m by using an online imputation method (Algorithm 3).

Algorithm 1. Ignoring the observations with missing values.

Require:
1: a data stream (DS)
2: a feature to predict (F_c), a feature with possible missing values (F_m)

Ensure:
3: $modelTree \leftarrow$ initialize the model tree to be trained using the data stream DS to predict the value of F_c
4: **while** data stream DS not finished **do**
5:    $OBS \leftarrow$ get the next observation of the data stream DS
6:    **if** OBS does not contain missing values for F_m **then**
7:       estimate error of $modelTree$ to predict the value of F_c in OBS
8:       train $modelTree$ with OBS
9:    **end if**
10: **end while**

Algorithm 2. Estimating the missing values with the mean.

Require:
1: a data stream (DS)
2: a feature to predict (F_c), a feature with possible missing values (F_m)

Ensure:
3: $modelTree \leftarrow$ initialize the model tree to be trained using the data stream DS to predict the value of F_c
4: $mean \leftarrow 0$
5: $count \leftarrow 0$
6: **while** data stream DS not finished **do**
7:    $OBS \leftarrow$ get the next observation of the data stream DS
8:    **if** OBS does not contain missing values for F_m **then**
9:       $OBS' \leftarrow$ fill OBS with $mean$
10:      estimate error of $modelTree$ to predict the value of F_c in OBS'
11:      train $modelTree$ with OBS'
12:    **else**
13:      estimate error of $modelTree$ to predict the value of F_c in OBS
14:      train $modelTree$ with OBS
15:      $currentValue \leftarrow$ current value of F_m
16:      $mean \leftarrow mean * (count - 1) + currentValue$
17:      $count \leftarrow count + 1$
18:    **end if**
19: **end while**

As these strategies don't offer guarantee about the accuracy of the obtained trained model tree, we propose an approach to adjust the estimated values for missing data in order to have a positive impact on the learned model tree (Algorithm 4). The suggested method aims at choosing a new estimation for each missing value by using a range defined with: (a) the value that is estimated by a given imputation method. (b) the current uncertainty/error of the imputation method.

Algorithm 3. Estimating the missing values with an imputation method.

Require:
 1: a data stream (DS)
 2: a feature to predict (F_c), a feature with possible missing values (F_m)
Ensure:
 3: $modelTree$ ← initialize the model tree to be trained using the data stream DS to predict the value of F_c
 4: $imputationMethod$ ← initialize an imputation method for estimating missing values of F_m
 5: **while** data stream DS not finished **do**
 6:    OBS ← get the next observation of the data stream DS
 7:    **if** OBS contains missing values for F_m **then**
 8:       $ESTIM$ ←estimate the missing value for F_m by using $imputationMethod$
 9:       OBS' ←fill OBS with $ESTIM$
10:       estimate error of $modelTree$ to predict the value of F_c in OBS'
11:       train $modelTree$ with OBS'
12:    **else**
13:       train $imputationMethod$ with OBS
14:       estimate error of $modelTree$ to predict the value of F_c in OBS
15:       train $modelTree$ with OBS
16:    **end if**
17: **end while**

First of all, an initial model tree and an imputation method are initialized (Algorithm 4 - lines 3, 4). Then, while the data stream is not finished (Algorithm 4 - line 5), the following steps are repeated:

- The next observation of the stream is considered (Algorithm 4 - line 6). If some values are missing from the considered observation then:
  - An estimation is computed for the missing value by applying the imputation method (Algorithm 4 - line 8).
  - The confidence level of the imputation method is evaluated by the algorithm using the Mean Absolute Error (Algorithm 4 - line 9).
  - By defining a boundary with this confidence level, the algorithm tries different estimations in such a way that the selected estimation will have a low impact on the trained model tree (Algorithm 4 - lines 10, 11, 12). This step is *time-efficient*, because it consists in simply selecting the estimation which does not tend to increase the model tree's error-rate.
  - The selected estimation is used to fill the incomplete observation (Algorithm 4 - line 15). This completed observation is used to train the model tree (Algorithm 4 - line 17).
- If the considered observation is complete then it is used to train both the imputation method (Algorithm 4 - line 19) and the model tree (Algorithm 4 - line 21).

Progressively, by incrementally processing the data stream, the algorithm aims at training the predictive model tree with adjusted estimations for the

---

Algorithm 4. Estimating the missing values with an imputation model tree, and adjusting the estimations.

Require:
 1: a data stream (DS)
 2: a feature to predict (F_c), a feature with possible missing values (F_m)
Ensure:
 3: $modelTree \leftarrow$ initialize the model tree to be trained using the data stream DS to predict the value of F_c
 4: $imputationMethod \leftarrow$ initialize an imputation method for estimating missing values of F_m
 5: **while** data stream DS not finished **do**
 6:   $OBS \leftarrow$ get the next observation of the data stream DS
 7:   **if** OBS contains missing values for F_m **then**
 8:     $ESTIM \leftarrow$ estimate the missing values for F_m by using $imputationMethod$
 9:     $MAE \leftarrow$ evaluate the current Mean Absolute Error of $imputationMethod$
10:     **for** VAL between $[ESTIM - MAE, ESTIM + MAE]$ **do**
11:       $OBS_{val} \leftarrow$ fill OBS with VAL
12:       $impact(VAL) \leftarrow$ measure the impact of training $modelTree$ with OBS_{val}
13:     **end for**
14:     select VAL_{best} for which $impact(VAL_{best})$ is lower
15:     $OBS' \leftarrow$ fill OBS with VAL_{best}
16:     estimate error of $modelTree$ to predict the value of F_c in OBS
17:     train $modelTree$ with OBS'
18:   **else**
19:     train $imputationMethod$ with OBS
20:     estimate error of $modelTree$ to predict the value of F_c in OBS
21:     train $modelTree$ with OBS
22:   **end if**
23: **end while**

---

incomplete data. For each missing value in the stream, it tends to favor an estimation that leads to a more accurate and more interpretable model tree, by potentially neglecting the quality of the estimation.

The algorithm do not need specific settings, so the user only has to select a stream, a continuous class attribute and an attribute with potential missing values. Nevertheless, integrating additional parameters is possible. For example, during the preliminary tests, we have tried to add a parameter to set the size of the computed boundaries (Alg.), but we have quickly seen it was useless. In addition, the computed boundary may be obtained by using other statistics about the considered feature (standard deviation, for example).

4 Experiments

4.1 Prototype

In order to validate the approach described in this paper, a prototype has been implemented as a JAVA standalone tool. It is mainly based on Cadral, an in-house data analysis platform [6] leveraging MOA, a widely-used data mining library for data streams [2]. More precisely, MOA provides algorithms for model tree induction on streams, and especially an implementation of the FIMT-DD algorithm [17].

The prototype provides a framework to dynamically analyze data streams by allowing a high user interactivity (play/pause/stop the stream, select the processing speed, process by bulk, etc.). To this end, it provides a user interface for data stream exploration, including statistics and data visualizations (Fig. 2): the graphical representation of the model trees relies on the JUNG library [25] (Fig. 1), and the plots are made by using JFreeChart [13].

Various streams can be loaded and processed in the prototype because it is capable of importing/exporting data as CSV files or ARFF files (the format adopted by MOA and Weka [41]).

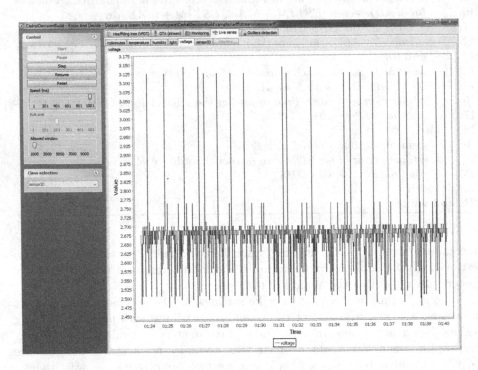

Fig. 2. The interface of the JAVA prototype helps to visually inspect data streams, for example the *voltage* values in the *sensor* data stream [1].

4.2 Evaluation Protocol

We have evaluated our approach on streams related to various domains (Tables 2 and 3). The data come from the *UCI Machine Learning repository* [1], the

Table 2. The considered data streams and their characteristics like the rows and features counts.

Data stream	Source repository	# rows	# considered features
Ailerons	Regression Datasets	13 750	41
Pole	Regression Datasets	15 000	48
MV Artificial Domain	Regression Datasets	40 768	11
Hyper Plane Stream	Stream Data Mining	100 000	11
KDD Cup 99	Stream Data Mining	145 585	42
3D spatial network	UCI Machine Learning	434 874	4
YearPredictionMSD	UCI Machine Learning	515 345	13
Forest Covertype	UCI Machine Learning	581 012	55
Sensor Stream	Stream Data Mining	2 219 803	6

Table 3. For each considered data stream, the considered continuous class to predict with the model tree, and the attribute used to create artificial missing values.

Data stream	Attribute to predict (class)	Attribute with missing values
Ailerons	Goal	Se
Pole	foo	f6
MV Artificial Domain	y	x10
Hyper Plane Stream	Attribute1	Attribute0
KDD Cup 99	dst host rerror rate	dst host srv rerror rate
3D spatial network	Altitude	Latitude
YearPredictionMSD	Year	attr11
Forest Covertype	Aspect	Elevation
Sensor Stream	Voltage	Humidity

Stream Data Mining Repository[1] and the *Regression Datasets repository*[2]. In order to realize meaningful tests, the selection was made in view to process a large range of rows counts and features count for the datasets.

To this end, these datasets have been considered as streams, i.e. they have been iteratively processed in an online way (in one pass, without accessing to the historical data). In each case:

(a) A continuous class has been considered to build model trees (Table 3).
(b) For a randomly selected attribute of the stream, artificial missing values have been introduced into 20 % of observations, in order to check the imputation method (Table 3).

---

[1] http://www.cse.fau.edu/xqzhu/stream.html.
[2] http://www.dcc.fc.up.pt/ltorgo/.

(c) 10 % of the observations have been used to compute the prediction error of the trained model tree (validation set). In this set, the real values are known for the selected continuous class and no missing data have been introduced.

Then, the different approaches to train the model tree have been tested on these streams (Algorithms 1, 2, 3 and 4). In each case, the following metrics have been measured to evaluate both the leading model tree and the results of the missing values imputation process:

(a) The model trees size after the training phase (obtained by simply counting the nodes and the leaves).
(b) The model trees accuracy regarding the validation set (MAE and RMSE, obtained by comparing the predicted values with the expected values).
(c) The confidence level of the missing value imputation (MAE and RMSE, obtained by comparing the estimated values with the original values – i.e. the values before creating *artificial* gaps into the the data streams).

Finally, the MOA's implementation of the FIMT-DD algorithm was configured with two parameters: *splitConfidence* and *gracePeriod*. Even if default values are provided by the implementation, we have realized a empirical sensitivity analysis in order to find the best configuration (i.e. leading to a good tradeoff for the accuracy and the size of the produced model tree): (a) $splitConfidence = 0.1$ (b) $gracePeriod = 200$.

4.3   Results

After the experiments, we have analyzed the model trees obtained after having applied different strategies to deal with missing values (Tables 4, 5, 6 and 7).

Firstly, we can observe that skipping the incomplete observations (Algorithm 1) leads in general to better results than training the model tree with observations that are filled with tested imputation methods (Algorithms 2, 3 and 4). For example, by considering the *YearPredictionMSD* data stream, the first one leads to a model tree with $RMSE = 10.55$ and the second one leads to a model tree with $RMSE = 61.48$. These results confirm that using an imputation method can have dramatic effects on the learned model tree.

Secondly, if the imputation is really needed (for values that are *missing not at random*), we can observe that our approach (Algorithm 4) generally leads to more accurate model trees, in comparison to those that are obtained by using the other approaches (Algorithm 2 and 3). For example, by considering the *KDD Cup 99* data stream, the *skipping approach* (Algorithm 1) leads to a model tree with $RMSE = 81.50$ and the second one leads to a model tree with $RMSE = 63.45$. We can note an exception for the *Forest Covertype* data stream: the accuracy is exactly the same for two approaches ($RMSE = 0.12$), but the model tree size is slightly smaller in the second case (5797 instead of 5805).

Thirdly, our technique has a positive impact on the model tree size too if we compare to the classical imputation method (Algorithm 3): for instance, the gain on the size is 28 % for the *Pole* stream, 4 % for the *KDD Cup 99* stream, etc.

Table 4. Results with the *skipping* method (Algorithm 1). For each data stream, the size and the error rate of the trained model tree are reported (MAE and RMSE are evaluated on the validation set, i.e. 10 % of the values).

Data stream	Trained model tree		
	Model tree size	MAE	RMSE
Ailerons	83	± 0.00	± 0.00
Pole	75	± 8.68	± 14.50
MV Artificial Domain	221	± 1.28	± 1.70
Hyper Plane Stream	475	± 0.50	± 0.58
KDD Cup 99	1 131	± 0.75	± 81.50
3D spatial network	3 403	± 12.85	± 16.33
YearPredictionMSD	4 091	± 7.91	± 10.55
Forest Covertype	4 641	± 0.07	± 0.12
Sensor Stream	17 737	± 0.06	± 0.14

Table 5. Results with the *mean* estimation method (Algorithm 2). For each data stream, the size and the error rate of the trained model tree are reported (MAE and RMSE are evaluated on the validation set, i.e. 10 % of the values). Moreover, the error rates of the missing values imputation are reported too (MAE and RMSE are evaluated on the fake missing data, i.e. 20 % of data).

Data stream	Trained model tree			Missing values imputation		
	Model tree size	MAE	RMSE	MAE	RMSE
Ailerons	73	± 0.00	± 0.00	± 0.00	± 0.00
Pole	107	± 8.68	± 14.50	± 17.05	± 20.93
MV Artificial Domain	287	± 1.23	± 1.62	± 50.45	± 58.16
Hyper Plane Stream	655	± 0.50	± 0.58	± 0.50	± 0.57
KDD Cup 99	1 423	± 0.60	± 63.45	± 0.13	± 0.32
3D spatial network	4 273	± 12.91	± 16.36	± 0.24	± 0.28
YearPredictionMSD	5 123	± 7.90	± 10.55	± 6.28	± 8.32
Forest Covertype	5 805	± 0.08	± 0.12	± 0.14	± 0.18
Sensor Stream	22 177	± 0.06	± 0.14	± 4.91	± 15.41

It is interesting because it is not the main goal of our technique. But again, skipping the incomplete observations provides model trees that are smaller than the other approaches, included our approach. And for some cases (the *Sensor* stream for example), the obtained model trees are big so it does not help their visualization/interpretation.

Fourthly, if we compare our approach (Algorithm 4) to the classical imputation approach (Algorithm 3), we can see that the missing values imputation is positively impacted. For instance, if we consider the *Sensor* stream, the confidence

Table 6. Results with the *model tree* estimation method (Algorithm 3). For each data stream, the size and the error rate of the trained model tree are reported (MAE and RMSE are evaluated on the validation set, i.e. 10 % of the values). Moreover, the error rates of the missing values imputation are reported too (MAE and RMSE are evaluated on the fake missing data, i.e. 20 % of data).

Data stream	Trained model tree			Missing values imputation	
	Model tree size	MAE	RMSE	MAE	RMSE
Ailerons	105	± 0.02	± 0.19	± 0.01	± 0.01
Pole	139	± 28.73	± 32.74	± 7.09	± 13.14
MV Artificial Domain	369	± 26.87	± 33.00	± 51.14	± 59.17
Hyper Plane Stream	661	± 0.50	± 0.58	± 0.51	± 0.59
KDD Cup 99	1 423	± 1.01	± 108.58	± 0.14	± 0.34
3D spatial network	4 273	± 13.28	± 16.72	± 0.20	± 0.23
YearPredictionMSD	5 123	± 43.53	± 61.48	± 5.55	± 7.30
Forest Covertype	5 797	± 0.08	± 0.12	± 0.09	± 0.14
Sensor Stream	22 177	± 1.15	± 1.52	± 4.23	± 15.40

Table 7. Results with our *adjustment* method (Algorithm 4). For each data stream, the size and the error rate of the trained model tree are reported (MAE and RMSE are evaluated on the validation set, i.e. 10 % of the values). Moreover, the error rates of the missing values imputation are reported too (MAE and RMSE are evaluated on the fake missing data, i.e. 20 % of data).

Data stream	Trained model tree			Missing values imputation	
	Model tree size	MAE	RMSE	MAE	RMSE
Ailerons	95	± 0.01	± 0.01	± 0.00	± 0.01
Pole	101	± 7.97	± 13.35	± 3.80	± 10.83
MV Artificial Domain	287	± 1.21	± 1.60	± 14.46	± 22.52
Hyper Plane Stream	637	± 0.50	± 0.58	± 0.18	± 0.34
KDD Cup 99	1 423	± 0.60	± 63.45	± 0.14	± 0.31
3D spatial network	4 273	± 12.90	± 16.36	± 0.05	± 0.10
YearPredictionMSD	5 123	± 7.91	± 10.55	± 1.95	± 4.05
Forest Covertype	5 805	± 0.08	± 0.12	± 0.05	± 0.09
Sensor Stream	22 177	± 0.06	± 0.14	± 1.68	± 14.80

level of the imputation is better by using our approach ($RMSE = 14.8$) than by using the other one ($RMSE = 15.4$).

Fifthly, we can say that the *mean* estimation is surprisingly not so bad: in our tests, replacing missing values by the current mean instead of estimating them with an imputation method leads to more or less the same accuracy for the leading model tree.

As a conclusion, we can say that it is globally more efficient to ignore incomplete observations during the training of a model tree from a stream. In fact, it saves a computation overhead and it avoids to introduce uncertain estimations in the model tree. This result was already observed for he induction of classification methods on classical datasets [30].

Nevertheless, it is not always desirable to skip incomplete parts because it can introduce a bias (in particular when data are not missing at random, for example). In this case, the presented results show that our adaptation approach (Algorithm 4) allows to obtain results which are better than using a mean estimation (Algorithm 2) or a more sophisticated imputation (Algorithm 3).

5   Conclusion

In this paper, we studied the impact of several missing values estimation methods to build predictive model trees on incomplete data streams. Moreover, we presented an algorithm that aims at adjusting the missing values estimation in order to help the training phase in a way that resulting model trees are more accurate. Our method helps to obtain model trees that are more correct than those obtained with other imputation methods; it is helpful when missing data estimation is really needed, especially when values are not missing at random.

The considered imputation methods were integrated and tested in a JAVA prototype, and the effectiveness of our algorithm was demonstrated and discussed on various data streams.

In future works, we will apply our method on large real-world data streams related to e-commerce and live sensors management. In addition, we have in view to improve the estimation method by applying dynamically an appropriate strategy according to the randomness of the missing data.

Acknowledgements. The project is supported by a grant from the Ministry of Economy and External Trade, Grand-Duchy of Luxembourg, under the RDI Law. Moreover, this work has been realized in partnership with the infinAIt Solutions S.A. company (http://infinait.eu), so we would like to thank Gero Vierke and Helmut Rieder for their help.

References

1. Bache, K., Lichman, M.: UCI Machine Learning Repository (2013)
2. Bifet, A., Holmes, G., Kirkby, R., Pfahringer, B.: MOA: Massive Online Analysis. J. Mach. Learn. Res. **11**, 1601–1604 (2010)
3. Breiman, L., et al.: Classification and Regression Trees. Chapman & Hall, New York (1984)
4. Breslow, L.A., Aha, D.W.: Simplifying decision trees: a survey. Knowl. Eng. Rev. **12**(1), 1–40 (1997)
5. Cortez, P., Cerdeira, A., Almeida, F., Matos, T., Reis, J.: Modeling wine preferences by data mining from physicochemical properties. Decis. Support Syst. **47**(4), 547–553 (2009). Smart Business Networks: Concepts and Empirical Evidence

6. Didry, Y., Parisot, O., Tamisier, T.: Engineering data intensive applications with cadral. In: Luo, Y. (ed.) CDVE 2015. LNCS, vol. 9320, pp. 28–35. Springer, Heidelberg (2015). doi:10.1007/978-3-319-24132-6_4
7. Domingos, P., Hulten, G.: Mining high-speed data streams. In: Proceedings of the Sixth ACM SIGKDD International Conference on Knowledge Discovery and Data Mining, pp. 71–80. ACM (2000)
8. Enders, C.K.: Applied Missing Data Analysis. Guilford Publications, New York (2010)
9. Farhangfar, A., Kurgan, L., Dy, J.: Impact of imputation of missing values on classification error for discrete data. Pattern Recogn. 41(12), 3692–3705 (2008)
10. Féraud, R., Clérot, F.: A methodology to explain neural network classification. Neural Networks 15(2), 237–246 (2002)
11. Fong, S., Yang, H.: The six technical gaps between intelligent applications, real-time data mining: a critical review. J. Emerg. Technol. Web Intell. 3(2), 63–73 (2011)
12. Frank, E., Mayo, M., Kramer, S.: Alternating model trees. In: 30th Annual ACM Symposium on Applied Computing, SAC 2015, pp. 871–878. ACM, NY (2015)
13. Gilbert, D.: The jfreechart class library: Developer Guide. Object Refinery 7 (2002)
14. Hang, Y., Fong, S.: An experimental comparison of decision trees in traditional data mining and data stream mining. In: 6th International Conference on Advanced Information Management and Service (IMS), pp. 442–447. IEEE (2010)
15. Hoeffding, W.: Probability inequalities for sums of bounded random variables. J. Am. Stat. Assoc. 58(301), 13–30 (1963)
16. Ikonomovska, E., Gama, J.: Learning model trees from data streams. In: Boulicaut, J.-F., Berthold, M.R., Horváth, T. (eds.) DS 2008. LNCS (LNAI), vol. 5255, pp. 52–63. Springer, Heidelberg (2008)
17. Ikonomovska, E., Gama, J., Džeroski, S.: Learning model trees from evolving data streams. Data Min. Knowl. Discov. 23(1), 128–168 (2011)
18. Ikonomovska, E., Gama, J., Sebastião, R., Gjorgjevik, D.: Regression trees from data streams with drift detection. In: Gama, J., Costa, V.S., Jorge, A.M., Brazdil, P.B. (eds.) DS 2009. LNCS, vol. 5808, pp. 121–135. Springer, Heidelberg (2009)
19. Junninen, H., Niska, H., Tuppurainen, K., Ruuskanen, J., Kolehmainen, M.: Methods for imputation of missing values in air quality data sets. Atmos. Environ. 38(18), 2895–2907 (2004)
20. Kotsiantis, S.B.: Decision trees: a recent overview. Artif. Intell. Rev. 39(4), 261–283 (2013)
21. Marwala, T., IGI Global: Computational intelligence for missing data imputation, estimation and management: knowledge optimization techniques. Information Science Reference, Herhsey (2009)
22. Muñoz, J., Felicísimo, Á.M.: Comparison of statistical methods commonly used in predictive modelling. J. Veg. Sci. 15(2), 285–292 (2004)
23. Murthy, S.K.: Automatic construction of decision trees from data: a multi-disciplinary survey. Data Min. Knowl. Discov. 2(4), 345–389 (1998)
24. Mwale, F.D., Adeloye, A.J., Rustum, R.: Infilling of missing rainfall and streamflow data in the Shire River basin, Malawi-a SOM approach. Phys. Chem. Earth 50, 34–43 (2012)
25. O'Madadhain, J., Fisher, D., White, S., Boey, Y.: The JUNG (Java Universal Network/Graph) framework. Technical report, UCI-ICS (2003)
26. Patel, K., Mehta, R.G., Raghuvanshi, M.M., Vadnere, N.N.: Incremental missing value replacement techniques for stream data. Int. J. Comput. Appl. 122(17), 9–13 (2015)

27. Pham, N.-K., Do, T.-N., Poulet, F., Morin, A.: Treeview, exploration interactive des arbres de decision. Revue d'Intelligence Artificielle **22**(3–4), 473–487 (2008)
28. Quinlan, J.R.: Learning with continuous classes. In: 5th Australian joint Conference on Artificial Intelligence, vol. 92, pp. 343–348, Singapore (1992)
29. Rubin, D.B.: Inference and missing data. Biometrika **63**(3), 581–592 (1976)
30. Saar-Tsechansky, M., Provost, F.: Handling missing values when applying classification models (2007)
31. Shmueli, G., Koppius, O.R.: Predictive analytics in information systems research. Mis Q. **35**(3), 553–572 (2011)
32. Siegel, E.V.: Competitively evolving decision trees against fixed training cases for natural language processing. Adv. Genet. Program. **19**, 409–423 (1994)
33. Smith, J.D., Borckardt, J.J., Nash, M.R.: Inferential precision in single-case time-series data streams: how well does the em procedure perform when missing observations occur in autocorrelated data? Behav. Ther. **43**(3), 679–685 (2012)
34. Stiglic, G., Kocbek, S., Pernek, I., Kokol, P.: Comprehensive decision tree models in bioinformatics. PLoS ONE **7**(3), e33812 (2012)
35. Tfwala, S.S., Wang, Y.-M., Lin, Y.-C.: Prediction of missing flow records using multilayer perceptron and coactive neurofuzzy inference system. Sci. World J. (2013)
36. Tran, T.T., Peng, L., Diao, Y., McGregor, A., Liu, A.: Claro: modeling and processing uncertain data streams. VLDB J. Int. J. Very Large Data Bases **21**(5), 651–676 (2012)
37. Buuren, S.V.: Flexible Imputation of Missing Data. CRC Press, Boca Raton (2012)
38. Hulse, J.V., Khoshgoftaar, T.M.: A comprehensive empirical evaluation of missing value imputation in noisy software measurement data. J. Syst. Softw. **81**(5), 691–708 (2008)
39. Walters, D.K.W., Linn, R.T., Kulas, M., Cuddihy, E., Chonghua, W., Granger, C.V.: Selecting modeling techniques for outcome prediction: Comparison of artificial neural networks, classification and regression trees, and linear regression analysis for predicting medical rehabilitation outcomes. J. Am. Med. Inform. Assoc. Suppl. S, vol. 1187 (1999)
40. Wang, Y., Witten, I.H.: Induction of model trees for predicting continuous classes (1996)
41. Witten, I.H., Frank, E., Hall, M.A.: Data Mining: Practical Machine Learning Tools and Techniques. Elsevier, San Francisco (2011)
42. Zhang, P., Zhu, X., Shi, Y., Guo, L., Xindong, W.: Robust ensemble learning for mining noisy data streams. Decis. Support Syst. **50**(2), 469–479 (2011)
43. Zhu, X., Xindong, W.: Class noise vs. attribute noise: a quantitative study. Artif. Intell. Rev. **22**(3), 177–210 (2004)
44. Zhu, X., Zhang, P., Wu, X., He, D., Zhang, C., Shi, Y.: Cleansing noisy data streams. In: ICDM 2008, pp. 1139–1144. IEEE (2008)
45. Žliobaitė, I., Hollmén, J.: Optimizing regression models for data streams with missing values. Mach. Learn. **99**(1), 47–73 (2015)

Hypergraph-Based Access Control Using Organizational Models and Formal Language Expressions – $\mathcal{HGAC}$

Alexander Lawall[✉]

Institute for Information Systems, Hof University, Alfons-Goppel-Platz 1,
95028 Hof, Germany
alexander.lawall@gmail.com

Abstract. In all organizations, access assignments are essential in order
to ensure data privacy, permission levels and the correct assignment of
tasks. Traditionally, such assignments are based on total enumeration,
with the consequence that constant effort has to be put into maintaining the assignments. This problem still persists when using abstraction
layers, such as group and role concepts, e.g. Access Control Matrix and
Role-Based Access Control. Role and group memberships are statically
defined and members have to be added and removed constantly. This
paper describes a novel approach – Hypergraph-Based Access Control
$\mathcal{HGAC}$ – to assign human and automatic subjects to access rights in a
declarative manner. The approach is based on an organizational metamodel and a declarative language. The language is used to express queries
and formulate predicates. Queries define sets of subjects based on their
properties and their position in the organizational model. They also contain additional information that causes organization-specific and user-
defined relations to be active or inactive depending on predicates. In
$\mathcal{HGAC}$, the subjects that have a specific permission are determined by
such a query. The query itself is not defined statically but created by traversing a hypergraph path. This allows a structured aggregation of permissions on resources. Consequently, multiple resources can share parts
of their queries.

Keywords: Access control · Attribute-based access control ·
Language expressions · Organizational metamodel · Organizational
model · Identity management

1 Introduction

Nowadays, companies have to deal with permanent change. Subjects (e.g. human
actors, machines, printers, etc.) join, leave or move within the organization. The
rearrangement of whole departments is also common. Therefore, the flexibility of
the organizational structure is essential to react to such changes, cf. [29]. Otherwise, the organization risks to lose their partners, e.g. deliverers and customers,
and elimination from the market, cf. [13].

© Springer International Publishing Switzerland 2016
M. Helfert et al. (Eds.): DATA 2015, CCIS 584, pp. 98–119, 2016.
DOI: 10.1007/978-3-319-30162-4_7

The organizational structure is currently shaped by work in project teams, global teams, networks and global teams in networks (cf. [13]). [20] formalizes a metamodel for modeling arbitrary organization structures that provides the required flexibility and complexity.

Access assignments have to be appropriate to policies that are declared in the company, cf. [7, p. 4]. Consequently, the validity of relations has to be restricted to realize these policies, cf. [22]. This is fulfilled by different types of predicates assigned to the relations.

Language expressions restrict the validity based on context information, parameters handed from application systems and/or attributes of subjects respectively resources in a company (cf. [19]). Figure 3 shows an example of a restricted relation based on parameters. The restricted relation with the language expression `damage > "1500"` is traversed if the parameter handed from the application system corresponding to "damage" fulfills the predicate.

A hyperedge in the organizational model – also called hyper-relation – restricts a relation by role. This means that a relation in the organizational model is only traversed if an entity acts in the appropriate role, cf. [19,22]. Figure 3 depicts a hyperedge between two subjects and a role. The relation is traversed if the origin subject acts in the role in that the hyperedge ends in.

This contribution compares approaches for defining access assignments and establishes the novel $\mathcal{HGAC}$. The conjunction of three components – organizational model, declarative language and hypergraph based access control – are a powerful mechanism to declare access assignments that are stable over time and fit the organizational circumstances and policies.

The formulated language expression (query) is used to define the access assignments in application systems. It defines policies of the company. Policies do not change often, and neither does the expression. Consequently, all application systems create no maintenance effort concerning access assignments after the initial definition of the expressions. Every time a subject joins, leaves or moves within the company, only logically centralized organizational model, cf. Sect. 6, has to be changed. The application systems are not affected.

Knorr describes in [12] an approach to assign access rights by workflows that are modeled with petri nets. If a subject is assigned to a task in a process, the subject automatically gets the rights to objects needed to execute the task. The definition of responsible subjects is done by role-based access control (RBAC). The approach is only valid in workflows. All other application systems are excluded[1].

Outline

In order to demonstrate key issues of widespread approaches, Sects. 2.1 and 2.2 describe the access control matrix and role-based access control.

Section 3 then introduces the novel approach called hypergraph-based access control. After the organizational metamodel (Sect. 3.1), an example of an organizational model is described (Sect. 3.2). Section 3.3 defines the formal specifi-

---

[1] The approach is only suited for process-oriented organizational structures. The functional and divisional perspective is omitted.

cation of $\mathcal{HGAC}$, which combines organizational information with access right assignment. This novel approach is explained by example in Sect. 3.4.

The subsequent Sect. 4 illustrates the proposed declarative language with syntax (Sect. 4.1) and semantics (Sect. 4.2). The language is used to declare the specific subjects that are assigned to access rights.

The paper concludes with a case study (Sect. 5) and the overall conclusion of the contribution (Sect. 6).

2   Access Control

For the definition of access rights exist different approaches. The following sections describe the access control matrix (ACM) and the wide-spread role-based access control (RBAC).

The basic model of access control consists generally of a tuple, cf. [1, p. 22] and [7, p. 6]. It consists of sets $\mathcal{S}, \mathcal{R}$ and $\mathcal{O}$. The set $\mathcal{S}$ includes all subjects substantiated by enumeration and represents users and processes. The access rights $\mathcal{R}$, e.g. *read, write* for files and *execute* for processes, are the operations on concrete objects of the set $\mathcal{O}$. The elements of $\mathcal{O}$ - files, processes, tables, devices, and so on - are the objects on which subjects have access rights.

> "There is usually a direct relationship between the cost of administration and the number of associations that must be managed in order to administer an access control policy: The larger the number of associations, the costlier and more error-prone access control administration."
> [4, p. 19]

A concrete access right $\mathcal{Z}$ is defined as $\mathcal{Z} = (s, r, o)$ with $s \in \mathcal{S}, r \in \mathcal{R}$ and $o \in \mathcal{O}$. In general, there are two variations to define access rights. All subjects have all access rights on all objects except rights that are explicitly revoked with tuples $\mathcal{Z}$. Another concept is that no subject has any access rights on any object until the access right is explicitly defined. The second case is the most used approach[2].

2.1   Access Control Matrix – ACM

The basic idea of the access control matrix was introduced in [10]. The formalization of [27, 28] is used to describe the access control matrix (ACM).

The configuration of a concrete access control matrix is defined with $ACM = (\mathcal{S}, \mathcal{R}, \mathcal{O}, (R^t)_{s \in \mathcal{S}, o \in \mathcal{O}})$[3]. The access control matrix $(R^t)_{s \in \mathcal{S}, o \in \mathcal{O}}$ consists of elements $R^t \subseteq \mathcal{R}$, where subjects $s \in \mathcal{S}$ are represented as rows and objects $o \in \mathcal{O}$ are represented as columns. An entry R^t in the matrix is the access right R^t of subject s to object o (see Fig. 1).

A configuration in an application system (e.g. workflow management systems, internet portals, database management systems, enterprise resource planning systems) with processes and files is given with:

---

[2] This case is used in the remaining paper.
[3] t indicates the point of time of a configuration in a system.

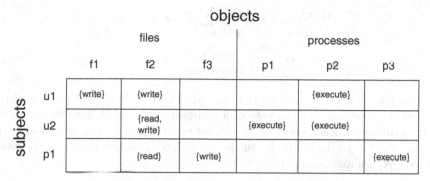

(a) Configuration of an Access Control Matrix (cf. [28]).

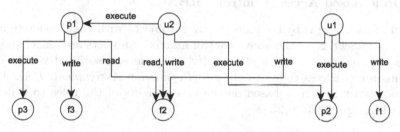

(b) Access Control Matrix as graph.

Fig. 1. Representations of an Access Control Matrix.

- $\mathcal{S} = \{u1, u2, p1\}$ is the set of users $u1, u2$ and process $p1$
- $\mathcal{R} = \{\text{read, write, execute}\}$ is the set of rights for processes (execute) and files (read, write)
- $\mathcal{O} = \{f1, f2, f3, p1, p2, p3\}$ is the set of objects with files and processes
- $(R^t)_{s \in \mathcal{S}, o \in \mathcal{O}}$ is represented as Fig. 1a respectively Fig. 1b

Discussion. The static definition of subjects assigned with access rights on objects is a problem regarding the continuous change in companies. Especially human subjects joining, moving or leaving companies are of interest concerning access rights and policies. Subjects are not limited to persons. Also automatic subjects like machines, computers and agents (cf. [18]) are involved. Software and hardware are similarly fluctuating in companies. The combination of the different application systems $\mathcal{K}$ and applications $\mathcal{A}$ makes the validity of access rights/policies at a specific point in time error-prone. A consistent state across all application systems is almost impossible.

Another aspect is the high maintenance effort resulting from continuous changes. The administrators, the responsible users or both are challenged by this effort.

The approximated maintenance effort in ACM for subject $s \in \mathcal{S}$ joining $j(s)$, moving $m(s)$ or leaving $l(s)$ is:

$$ACM_{j(s),m(s)} = \sum_{k\in\mathcal{K}}\sum_{a\in\mathcal{A}} |\mathcal{O}_{s,k,a}| \cdot |\mathcal{R}_{k,a}| \tag{1}$$

$$ACM_{l(s)} = \sum_{k\in\mathcal{K}}\sum_{a\in\mathcal{A}} |\mathcal{O}_{s,k,a}| \tag{2}$$

The determination of all subjects that are assigned to access right r on an object o is another problem. All subjects of a company have to be resolved from the involved column o. Each entry in the access control matrix has to be compared with the right r to decide if the subject is assigned to the right. In the worst case $|\mathcal{S}| \cdot |\mathcal{R}|$ comparisons per object are needed.

2.2 Role-Based Access Control – RBAC

Instead of assigning subjects individually to objects with their concrete access rights like in Sect. 2.1 – the access control matrix – subjects are associated with one or more roles (*User Assignment, UA*). A role is associated with a corresponding set of access rights to objects (*Permission Assignment, PA*). A subject's access to objects is based on the access rights of the roles to which the subject is assigned, cf. [6, 23–26].

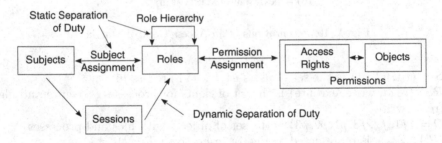

Fig. 2. Role-based Access Control adapted by [5].

Administrators for identity management tasks have to manage the access rights to an ideally small number of role definitions, rather than many individual user permissions, cf. [30].

There are more RBAC implementations extending the mentioned core RBAC. In [2,3], the extensions of RBAC include role hierarchies, constraints and the combination of role hierarchies and constraints (cf. Fig. 2). Role hierarchies are used to inherit access rights. For example, a head of a department is superior to his clerk and has also same access rights to all objects which the clerk is assigned to.

The constraint extension restricts the *Subject Assignment* and the assignment in role hierarchies with *Static Separation of Duty* in RBAC with constraints. *Dynamic Separation of Duty* restricts the active roles of a subject in a session, cf. Fig. 2.

Discussion. The main factor for using RBAC compared to the access control matrix is to reduce management costs. If access rights are assigned to a subject's role, the maintenance effort for managing individual access rights is eliminated, cf. [4, p. 19]. This means that as a subject moves into or out of a job function within an organization, access to the associated roles is granted and automatically rescinded. The administration effort is decreased because the reassignment of subjects to roles compared to the assignment of subjects to access rights has less work load. If there are more roles than subjects needed, the effort is higher.

The problem remains that the new role subject assignment has to be done in all application systems $\mathcal{K}$ and applications $\mathcal{A}$. Another aspect is that the *User Assignment* is static with regards to access rights. The access rights in a company for objects (e.g. processes, files,...) are often based on *context-*[4], *attribute-*[5] and/or *parameter-*[6]values. This is difficult with the RBAC approach. For each characteristic policy, a separate role is needed. Technical roles (roles in applications) are no longer job functions as intended in RBAC, cf. [4, p. 10]. Thus, the organizational job functions are not congruent to the roles in RBAC. An permanent effort in maintaining the mapping between job functions and roles is essential to ensure consistent policies/access rights. Thus a consistent access right assignment spread over all application systems with fluctuating subjects is hardly possible. The approximated maintenance effort in RBAC for subject $s \in \mathcal{S}$ joining $j(s)$, moving $m(s)$ or leaving $l(s)$ is:

$$RBAC_{j(s),m(s),l(s)} = \sum_{k \in \mathcal{K}} \sum_{a \subset \mathcal{A}} |Role_{s,k,a}| \tag{3}$$

The permission definition based on RBAC is more stable over time than using the access control matrix in the application systems respectively applications. Because the assignment of a role to permissions remains the same, the *Subject Assignment* changes.

Using RBAC, the determination of all subjects assigned to access right r on an object o is the following: Subjects are assigned to their roles by UA and permissions are assigned to roles by PA. In order to determine all assigned subjects, all roles with the right r on o need to be found by evaluating PA. In a second step, all subjects assigned to these roles have to be resolved by evaluating UA.

3   Hypergraph-Based Access Control – $\mathcal{HGAC}$

The $\mathcal{HGAC}$ approach is based on organizational information (cf. Sects. 3.1 and 3.2). This information is contained in a hypergraph, as defined in Sect. 3.3. Subjects that have access rights on objects are declared using language expressions (cf. Sect. 4).

---

[4] The context in which a subject acts (e.g. "purchase" in a workflow).

[5] Access rights are assigned using attributes of e.g. a subject (like "Hiring Year" > 2).

[6] Access rights defined by parameters passed from an application system. If for example the "damage" in an insurance case amounts to 200000, only subjects fulfilling this are responsible.

3.1   Organizational Metamodel

The organizational model, described in Sect. 3.2 conforms to the metamodel formulated in this section. The metamodel consists in general of the set of entity-types $\mathcal{E}$ and the relation-types $\mathcal{REL}$.

The excerpt of the metamodel[7] consists of the following entity-types $\mathcal{E} = \mathcal{OU} \cup \mathcal{RO} \cup \mathcal{S}$:

- Organizational units $\mathcal{OU} = \mathcal{OU}^i \cup \mathcal{OU}^e \cup \mathcal{OU}^\Upsilon$, e.g. departments, subdivisions, etc.
  - Internal organizational units $\mathcal{OU}^i$ are used in the example, cf. Fig. 3. These entity-types are used to represent organizational units within the organization.
  - External organizational units $\mathcal{OU}^e$ extend the organizational units. These entity-types are used to model organizational units from partner organizations. This makes it possible to model inter-organizational scenarios and to specify access rights across company borders.
  - Templates of organizational units $\mathcal{OU}^\Upsilon$ (general organizational units) are situated on the template level, cf. [21]. The templates define the general structure of similar organizational units. A template for a *Claims Department* generally consists of one *Head* and three *Clerks*. The departments *Car Damages* and *House Damages* are concrete manifestations of this template. In Fig. 3, the department *House Damages* deviates from the aforementioned template. It has an additional role *DB-Agent* and two subjects, instead of three, are assigned to the *Clerk* role.
- Roles $\mathcal{RO} = \mathcal{RO}^i \cup \mathcal{RO}^e \cup \mathcal{RO}^\Upsilon$ describe the roles in which subjects act.
  - Internal roles $\mathcal{RO}^i$ correspond to the internal organizational units. They represent the intra-organizational roles (cf. Fig. 3).
  - External roles $\mathcal{RO}^e$ are role entities from partner organizations.
  - Templates of roles $\mathcal{RO}^\Upsilon$ (general roles) are on the template level. Templates for roles are also part of the organizational metamodel, cf. [21]. Elements of $\mathcal{RO}^\Upsilon$ are used in conjunction with the templates of organizational units. Templates of roles include the number of subjects that can be assigned to a specific role (e.g. three subjects act as *Clerk*).
- Subjects $\mathcal{S} = \mathcal{S}^i \cup \mathcal{S}^e$ encompass human as well as automatic subjects.
  - Internal subjects $\mathcal{S}^i$ represent intra-organizational subjects (e.g. u_1, u_2, u_3, p_1 in Fig. 3).
  - External subjects $\mathcal{S}^e$ are used to specify subjects from partner organizations. Thus, access rights for external subjects can be declared.

The set of relation-types $\mathcal{REL} = \mathcal{REL}_s \cup \mathcal{REL}_o \cup \mathcal{REL}_u \cup \mathcal{REL}_e$ formally specifies the interconnections and consists of:

- The structural relation-type $rel_s \in \mathcal{REL}_s$ with

$$rel_s \subset (\mathcal{OU}^i \cup \mathcal{OU}^e) \times (\mathcal{OU}^i \cup \mathcal{OU}^e \cup \mathcal{RO}^i \cup \mathcal{RO}^e) \qquad (4)$$

---

[7] The basic metamodel is formally described in [14,19,21].

$$rel_s \subset (\mathcal{RO}^i \cup \mathcal{RO}^e) \times (\mathcal{RO}^i \cup \mathcal{RO}^e \cup \mathcal{S}) \tag{5}$$

$$rel_s \subset \mathcal{OU}^\Upsilon \times (\mathcal{OU}^\Upsilon \cup \mathcal{RO}^\Upsilon) \tag{6}$$

$$rel_s \subset \mathcal{RO}^\Upsilon \times \mathcal{RO}^\Upsilon \tag{7}$$

- The sets[8] of organization-specific relation-types $\mathcal{REL}_o$ (deputy, supervision, reporting) and user-defined relation-types $\mathcal{REL}_u$ with $\forall r \in \mathcal{REL}_o \cup \mathcal{REL}_u$:

$$r \subset (\mathcal{OU}^i \cup \mathcal{OU}^e) \times (\mathcal{OU}^i \cup \mathcal{OU}^e \cup \mathcal{RO}^i \cup \mathcal{RO}^e \cup \mathcal{S}) \tag{8}$$

$$r \subset (\mathcal{RO}^i \cup \mathcal{RO}^e) \times (\mathcal{RO}^i \cup \mathcal{RO}^e \cup \mathcal{S}) \tag{9}$$

$$r \subset \mathcal{S} \times (\mathcal{S} \cup \mathcal{RO}^i \cup \mathcal{RO}^e) \tag{10}$$

$$r \subset \mathcal{OU}^\Upsilon \times \mathcal{RO}^\Upsilon \tag{11}$$

$$r \subset \mathcal{RO}^\Upsilon \times \mathcal{RO}^\Upsilon \tag{12}$$

- The extension relation-type $rel_e \in \mathcal{REL}_e$ with

$$rel_e \subset \mathcal{OU}^\Upsilon \times (\mathcal{OU}^i \cup \mathcal{OU}^e) \tag{13}$$

$$rel_e \subset \mathcal{RO}^\Upsilon \times (\mathcal{RO}^i \cup \mathcal{RO}^e) \tag{14}$$

Formula (4) defines the structural interconnections between internal/external organizational units and roles. Formula (5) formalizes the interconnections between internal/external roles and other internal/external roles and subjects. The formulas (6) and (7) define the structural relations between templates of organizational units and templates of roles.

Organization-specific and user-defined relations are formalized in formulas (8) to (12). The semantics of the relation-types are different but some formulas are similar to the aforementioned. Formula (8) is almost identical to (4) but in (8) are the subjects included. (9) is syntactically identical to (5). Formula (10) interconnects subjects with other subjects and internal/external roles. The formulas (11) and (12) determine possible interconnections on the template level.

The remaining formulas (13) and (14) define connections between elements of the template level and their concrete representations on the instance level.

Role-dependent relations (cf. Fig. 3, deputyship between u_1 and u_2) are used to restrict organization-specific and user-defined relations. Whether the relations are active depends on the active role a subject acts in. Role-dependent relations can be used by relations following formulas (8) and (10). The tuple (x, y) with $x \in (\mathcal{RO}^i \cup \mathcal{RO}^e)$, $y \in (\mathcal{RO}^i \cup \mathcal{RO}^e \cup \mathcal{S})$ – corresponding to formula (9) – has the acting role already declared in variable x. Role-dependent relations do not exist on the template level, cf. formulas (11) and (12).

3.2 Organizational Model

For clarity, external and template entities (cf. Sect. 3.1) are excluded in this example. External entities are used analogously as internals. The manual modeling of the external entities is described in [19]. The automatic propagation of these entities to partner companies is described in [15]. The detailed mechanism of the templates are described in [21].

---

[8] The organization-specific and user-defined relations can be constrained, cf. [19].

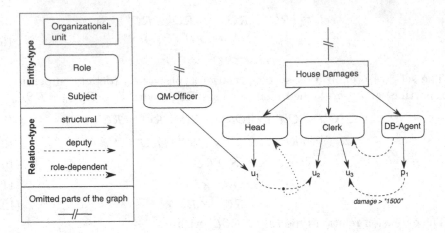

Fig. 3. Excerpt of an Organizational Model of an Insurance Company.

Structural Relations. Figure 3 shows an example model of an insurance company. The model consists of the department *House Damages*, the subjects u_1, u_2, u_3 (human subjects) and p_1 (automatic subjects) with their functional units *Head, Clerk* and *DB-Agent*. The subject u_1 is also working as *QM-Officer* a position within another department.

Organization-Specific Relations. Beside the structural relations, the company's model contains further relations – organization-specific relations[9] (e.g. deputy, supervisor and reporting relations) that interconnect entities. In this example, the *Head* u_1 has a deputy u_2 if u_1 acts as *Head* of the department *House Damages*[10]. If u_1 acts as *QM-Officer*, then u_2 is not a possible deputy.

In case the *DB-Agent* p_1 is unavailable, a constrained deputy relation to u_3 is evaluated at first. If the parameter-based[11] predicate (**damage > "1500"**) of the relation is fulfilled, it transfers all access rights of p_1 to u_3. In this scenario, the subject u_3 is a *Clerk* responsible for expensive insurance cases. If u_3 and p_1 are simultaneously unavailable, the general deputy relation between *DB-Agent* and *Clerk* is evaluated. The result set consists of subject u_2 which is then the deputy. u_3 is not an element of this set because of unavailability. This algorithm represents a prioritization mechanism.

Another variant to restrict the validity of relations are context-sensitive constraints (not included in the example). A constraint assigned to a deputy relation

---

[9] In [16–18], the complete metamodel and formal language including, i.a. constraints, are specified.

[10] It is possible to restrict any and all organization-specific relations to be role-dependent.

[11] If a value is handed from an application system to the organizational model via the formal language, the expression is called parameter-based.

of a subject is fulfilled if the context is identical to the context provided by an application system.

3.3   Formal Specification of $\mathcal{HGAC}$

The formal specification of a hypergraph defined by [9] will be redefined for the access control with $\mathcal{HGAC}$. A hypergraph as defined by [9] is a graph $G_{hyp} = (V, E)$ with the set of nodes $V = \{v_1, v_2, ..., v_n\}$ and the set of hyperedges $E = \{E_1, E_2, ..., E_m\}$. A directed hyperedge $E_i = (S, Z)$ consists of arbitrary non-empty sets of *start nodes* $S \subseteq V$ and *target nodes* $Z \subseteq V$.

The set of start nodes in $\mathcal{HGAC}$ contains one entity – the *Permission*. This start node is substituted in the formal specification as $\alpha := Permission$.

The access control hypergraph $G_{Perm} = (V, E)$ is defined as:

- The set of nodes $V = \{\alpha, o_1, ..., o_l\}$ with $o_g \in \mathcal{O} \land g = 1, ..., l$
  - The node α is the reference node of all access right definitions.
  - $\mathcal{O}$ the set of objects which are assigned to rights.
- The set of edges $E = \{r_j^h \mid r_j^h \in \xi_j \land h = 1, ..., d\}$ consists of ternary relations r_j^h (j : relation-types for rights, d : number of relations per relation-type)
  - $\mathcal{R}$: set of relation-types of rights
  - $\xi_j \in \mathcal{R}$: set of relations of a specific relation-type of a right
  - $\forall r_j^h (r_j^h \in \xi_j)$: $r_j^h \subseteq (\{\alpha\} \cup \xi_j) \times (\mathcal{O} \cup \xi_j) \times \mathcal{L}$:[12]
    * The element $\{\alpha\} \cup \xi_j$ of the tuple declares the start of an edge in G_{Perm}. $\alpha \in V$ is the first start node of every relation of a relation-type j of an hyperedge. Afterwards, arbitrary relations $r_j^b \in \xi_j$ with $b = 1, ..., |\xi_j| - 1$ can be the start of an edge of the same relation-type j.
    * $\mathcal{O} \cup \xi_j$ defines the end of an edge in the graph. This can be a concrete object $o \in (\mathcal{O} \subset V)$ or a set of relations of a relation-type ξ_j.
    * A language expression $\mathcal{L}$ is a valid element of language $L(G)$[13]. The empty word ε is included in the language $L(G)$ as well.
  - $f_{\mathcal{L}} : r_j^h \rightarrow \mathcal{L}$ extracts the language expression $\mathcal{L}$ assigned to the hyper-relation r_j^h.
- The subjects $\mathcal{S}$ result from evaluating the language expressions $\mathcal{L}$ on the model.
  - $\mathcal{L} \Rightarrow^* \mathcal{S}_{part} \subseteq \mathcal{S}$ defines the result set $\mathcal{S}_{part}$ of the language expression $\mathcal{L}$. The symbol $\Rightarrow^*$ indicates the resulting subjects of the language expression $\mathcal{L}$ by traversing the organizational model. The traversal algorithms are formalized in [19,21].
  - $\mathcal{S}_{path} = \bigcup_{p=1}^{e} \mathcal{S}_{part}$ with e equals the path length, defines the set of all subjects assigned to access right j on the path $P_j^{o \in \mathcal{O}}$. This path starts in α and ends in $o \in \mathcal{O}$. All language expressions $\mathcal{L}$ of the relations r_j^h on $P_j^{o \in \mathcal{O}}$

---

[12] The relation r_j^1 includes **always** the "Permission" (α) node.
[13] The syntax and semantic of the language is defined in [17].

are concatenated with OR[14]. The resulting expression is evaluated on the model to get the subjects $\mathcal{S}_{path}$.

The set of *all* subjects assigned to an access right j for an object $o \in \mathcal{O}$ can be evaluated differently. It is possible to compare all paths of a right j related to the object o.

Starting the traversal in object o is more efficient because an unnecessary evaluation of paths containing $\mathcal{O} \setminus o$ is excluded. The direction of the concatenation of the language expressions is in "reverse" order. The reverse and forward concatenation of language expressions results in identical subjects.

3.4 Definition of Access Rights

An example (cf. Fig. 4) configuration of the hypergraph $G_{Perm} = (V, E)$ is:

- $\mathcal{O} = \{f1, f2, f3, p1, p2, p3\}$: the set of objects containing files and processes
- $V = \{\alpha, f1, f2, f3, p1, p2, p3\}$
- $E = \{r_{read}^1, rel_s^1, rel_s^2, rel_s^3, rel_s^4, rel_s^5, r_{exec}^1, r_{exec}^2, r_{exec}^3, r_{exec}^4, r_{exec}^5\}$
  - $\mathcal{R} = \{READ, WRITE, EXECUTE\}$
  - $\xi_{read} = \{r_{read}^1\}$

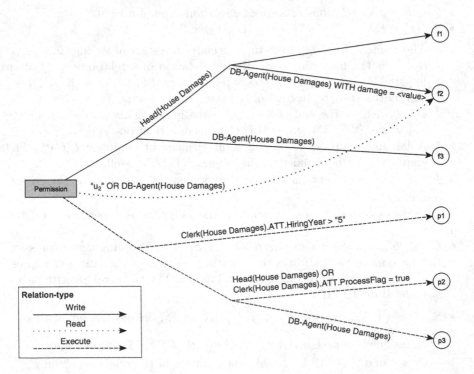

Fig. 4. Access Rights defined by using a Hypergraph and Formal Expressions.

---

[14] The empty word ε is excluded from this concatenation.

- $\xi_{write} = \{rel_s^1, rel_s^2, rel_s^3, rel_s^4, rel_s^5\}$
- $\xi_{exec} = \{r_{exec}^1, r_{exec}^2, r_{exec}^3, r_{exec}^4, r_{exec}^5\}$
- $r_{read}^1 = (\{\alpha\}, \{f2\}, $ "u2" OR DB-Agent(House Damages))
- $rel_s^1 = (\{\alpha\}, \{rel_s^2, rel_s^3\}, \varepsilon)$
- $rel_s^2 = (\{rel_s^1\}, \{f3\}, $ DB-Agent(House Damages))
- $rel_s^3 = (\{rel_s^1\}, \{rel_s^4, rel_s^5\}, $ Head(House Damages))
- $rel_s^4 = (\{rel_s^3\}, \{f2\}, $ DB-Agent(House Damages) WITH damage = "2000")
- $rel_s^5 = (\{rel_s^3\}, \{f1\}, \varepsilon)$
- $r_{exec}^1 = (\{\alpha\}, \{r_{exec}^2, r_{exec}^3\}, \varepsilon)$
- $r_{exec}^2 = (\{r_{exec}^1\}, \{p1\}, $ Clerk(House Damages).ATT.HiringYear > "5")
- $r_{exec}^3 = (\{r_{exec}^1\}, \{r_{exec}^4, r_{exec}^5\}, \varepsilon)$
- $r_{exec}^4 = (\{r_{exec}^3\}, \{p2\}, $ Head(House Damages) OR Clerk(House Damages).ATT.Processflag = "true")
- $r_{exec}^5 = (\{r_{exec}^3\}, \{p3\}, $ DB-Agent(House Damages))
- The set of subjects $\mathcal{S} = \{u_1, u_2, u_3, p_1\}$ (human and automatic)[15] (cf. Fig. 3):
  - $f_{\mathcal{L}}\left(r_{read}^1\right) = $ "u2" OR DB-Agent(House Damages) $\Rightarrow^* \mathcal{S}_{part} = \{u_2, p_1\}$
  - $f_{\mathcal{L}}\left(rel_s^2\right) = f_{\mathcal{L}}\left(r_{exec}^5\right) = $ DB-Agent(House Damages) $\Rightarrow^* \mathcal{S}_{part} = \{p_1\}$
  - $f_{\mathcal{L}}\left(rel_s^3\right) = $ Head(House Damages) $\Rightarrow^* \mathcal{S}_{part} = \{u_1\}$
  - $f_{\mathcal{L}}\left(rel_s^4\right) = $ DB-Agent(House Damages) WITH damage = "2000" $\Rightarrow^* \mathcal{S}_{part} = \{p_1\}$
  - $f_{\mathcal{L}}\left(r_{exec}^2\right) = $ Clerk(House Damages).ATT.HiringYear > "5" $\Rightarrow^* \mathcal{S}_{part} = \{u_2\}$[16]
  - $f_{\mathcal{L}}\left(r_{exec}^4\right) = $ Head(House Damages) OR Clerk(House Damages).ATT.Processflag = "true" $\Rightarrow^* \mathcal{S}_{part} = \{u_1, u_3\}$[17]
  - $f_{\mathcal{L}}\left(rel_s^1\right) = f_{\mathcal{L}}\left(rel_s^5\right) = f_{\mathcal{L}}\left(r_{exec}^1\right) = f_{\mathcal{L}}\left(r_{exec}^3\right) = \varepsilon \Rightarrow^* \mathcal{S}_{part} = \emptyset$
- The set of all subjects assigned to access rights on an object $o \in \mathcal{O}$:
  - Access right $\mathcal{WRITE}$ to an object $f1$ is assigned to subject u_1:
    * Path $P_{\mathcal{WRITE}}^{f1} = \left(rel_s^5, rel_s^3, rel_s^1\right)$
    * $\mathcal{S}_{path} = \emptyset \cup \{u_1\} \cup \emptyset = \{u_1\}$
  - Access right $\mathcal{WRITE}$ to an object $f2$ is assigned to subjects u_1, p_1 and access right $\mathcal{READ}$ to u_2, p_1:
    $\mathcal{WRITE}$:
    * Path $P_{\mathcal{WRITE}}^{f2} = \left(rel_s^4, rel_s^3, rel_s^1\right)$
    * $\mathcal{S}_{path} = \{p_1\} \cup \{u_1\} \cup \emptyset = \{u_1, p_1\}$
    $\mathcal{READ}$:
    * Path $P_{\mathcal{READ}}^{f2} = \left(r_{read}^1\right)$
    * $\mathcal{S}_{path} = \{u_2, p_1\}$
  - Access right $\mathcal{WRITE}$ to an object $f3$ is assigned to subject p_1:

---

[15] The following results can be different in cases of absence of subjects depending on deputy relations.

[16] u_3 is not included in the resulting set of subjects because the attribute-based predicate is not fulfilled.

[17] u_3 is included in the resulting set of subjects because the value of his attribute fulfills the attribute-based predicate.

    * Path $P_{WRITE}^{f3} = \left(rel_s^2, rel_s^1\right)$
    * $S_{path} = \{p_1\} \cup \emptyset = \{p_1\}$
- Access right $\mathcal{EXECUTE}$ to an object $p1$ is assigned to subject u_2:
    * Path $P_{\mathcal{EXECUTE}}^{p1} = \left(r_{exec}^2, r_{exec}^1\right)$
    * $S_{path} = \{u_2\} \cup \emptyset = \{u_2\}$
- Access right $\mathcal{EXECUTE}$ to an object $p2$ is assigned to subjects u_1, u_3:
    * Path $P_{\mathcal{EXECUTE}}^{p2} = \left(r_{exec}^4, r_{exec}^3, r_{exec}^1\right)$
    * $S_{path} = \{u_1, u_3\} \cup \emptyset \cup \emptyset = \{u_1, u_3\}$
- Access right $\mathcal{EXECUTE}$ to an object $p3$ is assigned to subject p_1:
    * Path $P_{\mathcal{EXECUTE}}^{p3} = \left(r_{exec}^5, r_{exec}^3, r_{exec}^1\right)$
    * $S_{path} = \{p_1\} \cup \emptyset \cup \emptyset = \{p_1\}$

The definition of a specific access right using (hyper-) relations and language expressions in conjunction with the organizational model simplifies the maintenance if the company's organization changes. This is established by the logically centralized organizational model and the language expressions that are stored in the hypergraph.

The maintenance effort for applying company policies decreases with $\mathcal{HGAC}$. In the example, the path $P_{WRITE}^{f2} = \left(rel_s^4, rel_s^3, rel_s^1\right)$ defines the access right $WRITE$ assigned to object $f2$. The subjects resulting from the language expression DB-Agent(House Damages) WITH damage = "2000" assigned to the relation rel_s^4 is exclusively valid for object $f2$.

The relation rel_s^3 with the language expression Head(House Damages) is valid for the objects $f1$ and $f2$. The expression does not have to be stored twice. Thus, $\mathcal{HGAC}$ provides a mechanism to avoid redundancies.

Access rights for new objects in application systems can be integrated at any point of the access right hypergraph. The user can reuse previously defined relations and assigned expressions. The user connects the new relation to the hypergraph. If needed, he assigns a new language expression to get the appropriate set of subjects that have access to the new object.

For example, a user adds an object $f4$ (e.g. financial report for the department House Damages) to an application system. The policy for this object implies that the Head and Clerks of the department House Damages can access it, but not the DB-Agent. Therefore, the user instantiates a relation of the type $WRITE$ and assigns the expression Clerk(House Damages). This new relation $rel_s^6 = (\{rel_s^3\}, \{f4\}, \text{Clerk(House Damages)})$ originates in rel_s^3[18] and ends in object $f4$. Thus, the policy is defined by reusing existing parts. This avoids redundancies.

For clarification, all redundant expressions (e.g. DB-Agent (House Damages)) shown in Fig. 4 are "redundantly" written to simplify reading. Such recurring expressions can be stored in macros, so that changes in the macro affect all (hyper-) relations that store this macro. Macros avoid redundancies in the definition of access rights respectively policies with $\mathcal{HGAC}$.

---

[18] rel_s^3 includes Head(House Damages).

4 Declarative Language

The domain-specific language $L(G)$ is defined by the context-free grammar $G = (N, \Sigma, P, S_G)$ with the set of non-terminals N, the alphabet Σ with $N \cap \Sigma = \emptyset$, the production rules P and the start symbol S_G with $S_G \in N$ (cf. [8]). Each production rule has the format $l \rightarrow r$ with $l \in N$ and $r \in (N \cup \Sigma)^*$.

4.1 Syntax

Non-terminals that expand only to one specific sequence of terminals (keywords) are represented as e.g. 'NOT', 'WITH'.

The grammar G_1 for defining *queries* is a tuple of:

- The set of non-terminals $N_1 = \{start, query, actor, funits, funit, oudef,$
  $ounits, ounit, relationTokens, withParams, contextDefinition,$
  $attConstraints, kcv, parameter, kvp, id, string\}$
- The alphabet of terminals $\Sigma_1 = \{$'a','b',...,'z','A','B',...,'Z','ä','ü','ö','Ä','Ü',
  'Ö','0','1',...,'9','_','-','(',')',',',',','*','=','<','>'$\}$[19]
- The set of production rules P_1[20]
  > $start \rightarrow query \mid query\ logic\ query \mid \varepsilon$
  > $query \rightarrow actor \mid actor\ \text{'AS'}\ funits$
  > $query \rightarrow query\ \text{'NOT'}\ query$
  > $query \rightarrow query\ \text{'FALLBACKTO'}\ query$
  > $query \rightarrow query\ \text{'WITH'}\ withParams$
  > $query \rightarrow funits\ \text{'('}oudef\text{')'}$
  > $query \rightarrow relationTokens\ \text{'('}query\text{')'}$
  > $query \rightarrow \text{'('}query\ logic\ query\text{')'}$
  > $query \rightarrow \text{'('}query\text{')'}.'attConstraints$
  > $actor \rightarrow \text{'*'} \mid id \mid string$
  > $funits \rightarrow funit \mid \text{'('}funit\ logic\ funit\text{')'}$
  > $funit \rightarrow \text{'*'} \mid id \mid string$
  > $oudef \rightarrow ounit \mid ounits\ logic\ ounits$
  > $ounits \rightarrow ounit \mid \text{'('}ounits\ logic\ ounits\text{')'}$
  > $ounit \rightarrow \text{'*'} \mid id \mid string \mid ounit\ \text{'SUBS'}$
  > $relationTokens \rightarrow (\text{'ALL'} \mid \text{'ANY'})?\ id\ (\text{'OF'} \mid \text{'TO'})$
  > $withParams \rightarrow contextDefinition \mid parameter \mid withParams\ ','$
  > $withParams$
  > $contextDefinition \rightarrow \text{'CONTEXT='}\ context\ (',' \ context)*$
  > $attConstraints \rightarrow \text{'ATT.'}\ kcv$
  > $kcv \rightarrow id\ comp\ string \mid \text{'('}kcv\ logic\ kcv\text{')'}$
  > $parameter \rightarrow kvp\ (',' \ kvp)*$
  > $kvp \rightarrow id\ \text{'='}\ string$
  > $logic \rightarrow \text{'AND'} \mid \text{'OR'}$
  > $comp \rightarrow (\text{'='} \mid \text{'<='} \mid \text{'>='} \mid \text{'<'} \mid \text{'>'} \mid \text{'!='})$

---

[19] The terminals derived from the non-terminal *string* are also included.

[20] Meaning of meta-symbols: ? means 0 or 1 and * means 0 to ∞ occurrences.

$id \rightarrow$ (['a'-'z','A'-'Z'] | '_' | 'Ä' | 'ä' | 'Ü' | 'ü' | 'Ö' | 'ö') (['a'-'z','A'-'Z'] | 'Ä' | 'ä' | 'Ü' | 'ü' | 'Ö' | 'ö' |[0 − 9]| '_' | '-')*

$string \rightarrow$ '"' id '"'

– The set of start symbols $S_{G_1} = \{start\}$

The grammar G_2 for defining *predicates on relations* is a tuple of:

– The set of non-terminals $N_2 = \{internal, relPred, parameteratt, context,$ $parameter, attribute, kcv, logic, id\}$
– The alphabet of terminals $\Sigma_2 = \Sigma_1$
– The set of production rules P_2[21]

$internal \rightarrow relPred| \ relPred \ logic \ relPred|\varepsilon$

$relPred \rightarrow context|parameteratt| \ '('relPred')' \ | \ '('relPred \ logicrel$ $Pred')'$

$parameteratt \rightarrow parameter|attribute| \ '('parameteratt \ logicparameter$ $att')'$

$context \rightarrow id| \ '('context \ logic \ context')'$

$parameter \rightarrow kcv| \ '('parameter \ logic \ parameter')'$

$attribute \rightarrow 'ATT.' \ kcv| \ '('attribute \ logic \ attribute')'$

– The set of start symbols $S_{G_2} = \{internal\}$

The grammar G is the result of the union of grammars G_1 and G_2[22]. This equals $G = \{N_1 \cup N_2, \Sigma_1, P_1 \cup P_2 \cup \{s_G \rightarrow start \mid internal\}, \{s_G\}\}$ A language expression $\mathcal{L}$ is syntactically correct if $\mathcal{L}$ is derivable starting from the set of start symbols of the language: $L(G) = \{\mathcal{L} \in \Sigma^* \mid S_G \leadsto_G^* \mathcal{L}\}$. The bottom-up approach for the syntactical evaluation is also possible.

4.2 Semantics

Language expressions are formulated *within* or *outside* of the organizational model. *Within* means that the expression is inside of an organizational model and represents a *predicate* assigned to a relation. The grammar for the definition of predicates is G_2.

The syntax for *queries* is defined by the grammar G_1. They are the *outside* perspective. Application systems pass language expressions to the organizational model (via organizational server, cf. Sect. 6, Fig. 6). The expression is then evaluated on the organizational model which yields subjects.

The semantics of the domain-specific language $L(G)$ is described informally for brevity. The semantics of *queries* that are formulated to get the appropriate set of subjects is the following[23] – $L(G_1)$:

---

[21] Production rules for non-terminals $kcv, logic$ and id correspond to those in P_1.

[22] Union of context-free grammars according to [11].

[23] Composite access rights can be defined by the concatenation of queries with logical AND and OR.

- **Based on Structural Relations** are queries describing a concrete subject, e.g. "u1", subjects having a specific role in a specific organizational unit e.g. Clerk (House Damages), or subjects having a specific role in any organizational unit e.g. Clerk(*).
- **Separation of Duty** is expressed by the NOT-clause mostly used in workflow management systems, e.g. SUPERVISOR OF(<initiator>) NOT <initiator>, to prevent the initiator of a process from approving his own request for purchase.
- **Prioritization** defines primary candidates and an alternative to fall back to. If the set of primary candidates is empty, the second query is evaluated, e.g. Clerk (House Damages) FALLBACKTO Head(House Damages).
- **Parameter Passing** is done using WITH. Application systems pass parameters to the organizational model, e.g. DB-Agent(House Damages) WITH damage ="1000".
- **Acting Role** describes the role in which a subject acts. This is used to decide if role-dependent relations are valid, e.g. the deputy relation between $u1$ and $u2$. This deputyship is only valid if $u1$ acts as $Head$ (cf. Sect. 3.2). There are different possibilities to formulate such queries. The implicit enumeration of roles e.g. Head(House Damages) specifies the role $Head$. The explicit variant to pass a role to the organizational model when directly naming a subject is possible with e.g. u1 AS Head.
- **Subjects Restricted by Attributes** are subjects that fulfill the attribute constraints. The expression Clerk(House Damages).ATT.HiringYear > "2" describes the subjects that have been on the job for more than two years.
- **Based on Organization-specific Relations** that are deputyship, supervision and reporting dependencies. This can also be used for audits. If, for example, all possible deputies have to be listed, the expression ANY DEPUTY OF(p1) can be used. The possible subjects are $u2$ and $u3$, stemming from the predicated and not predicated deputy relations from $p1$ and DB-$Agent$. ANY ignores predicates (e.g. damage > "1500" or role-dependent predicates). ALL indicates that the relation is to be followed transitively.

The semantic of *predicates* on relations that are used to declare organizational circumstances are[24] – $L(G_2)$:

- **General Validity** means that if no predicate is assigned to a relation (ε) it is always valid, e.g. the deputy relation between the roles DB-$Agent$ and $Clerk$.
- **Based on the Current Context the Subject is supposed to act in**, specific relations can be valid or invalid. This has consequences for organizational regulations, e.g. $u2$ can only be $u1$'s deputy, if $u1$ acts as $Head$, not if he acts as QM-$Officer$. Another context can be a "purchase". A predicate is e.g. purchase.damage >"1500". The deputyship between $p1$ and $u3$ is dependent on the context "purchase".
- **Based on Attributes of the Subjects**, subjects can be filtered from the result set. Relations on the path do not change their validity by this mechanism.

---

[24] Predicates can be based on a combination of context, attributes and parameters.

- **Based on Parameters** from application systems, the relations may be valid or invalid. This directly influences the traversal of relations within the organizational model.

Discussion. The novel approach $\mathcal{HGAC}$ reduces the maintenance effort to zero. If subjects join, move or leave the company, no changes have to be done concerning access right assignment. The access rights are immediately and automatically consistent.

On the one hand, this is enabled by the organizational (meta-) model of the company's circumstances, cf. Sect. 3.2, and on the other hand by the declarative language, cf. Sect. 4. The policies are formulated in language expressions that describe the requirements for access.

The expressions declare queries for policies/access rights in application systems. They are based on organizational structures (entities and relations) and consider properties of subjects (attributes).

Additionally, the language is used to define predicates. They are used for policies that are formulated on relations in the organizational model. The conjunction of queries and predicates is a powerful tool for defining policies.

Characteristic technical roles, as needed in RBAC, to define the needed policies/access rights are obsolete. Policies are described by language expressions and structured using hyperedges in $\mathcal{HGAC}$. Thus, changes – property changes (e.g. name, hiring year, salary, etc.) and relation changes (e.g. join, move, leave, new supervisor or deputy relation, etc.) – concerning subjects do not affect the access right and policy definitions.

If, for example, a subject $s \in \mathcal{S}$ joins $j(s)$, moves $m(s)$ or leaves $l(s)$ the company, the effort[25] in $\mathcal{HGAC}$ maintaining access rights in the application systems is:

$$\mathcal{HGAC}_{j(s),m(s),l(s)} = 0 \tag{15}$$

The only maintenance effort is in the organizational model, cf. Sect. 5.

Structuring access rights is limited in ACM and RBAC approaches. The access rights can be structured hierarchically, e.g. for directory rights and contained directories and files. This is an object-centered view. The objects are generally structured. $\mathcal{HGAC}$ favors an operator-centered view. The operators (i.a. read, write) are structured and the subjects are declared by the language expressions. This makes it easy to find all subjects that have a specific access right to a specific object.

5   Case Study

In order to validate and compare the approach, a case study was conducted. Two objects with access rights assigned using RBAC were evaluated. These objects are directories on a file server:

---

[25] The effort is identical for relation changes, e.g. adding a supervisor relation in the organizational model.

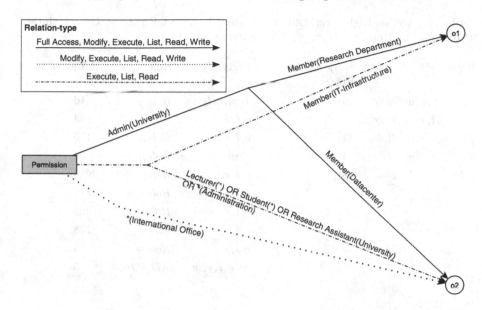

Fig. 5. Access Rights of the Case Study in $\mathcal{HGAC}$.

- "Project X" (cf. Table 1) is an arbitrarily selected directory of a research project.
- "International" (cf. Table 2) is a directory containing resources for academic international affairs.

Tables 1 and 2 show the permissions that are assigned to different roles for the objects. They also list the number of subjects that are assigned to the individual roles. The set $\mathcal{R} = \{full\ access, modify, execute, list, read, write\}$ defines the access rights that can be assigned. For brevity, they are denoted in the tables as $\{f, m, e, l, r, w\}$ correspondingly.

As can be seen from the amount of subjects, an ACM approach would not be practical with almost 4000 subjects, even for such a small number of objects[26].

A detailed look at the subject assignments revealed a number of inconsistencies:

- Subjects occur multiple times in different roles.
- Subjects are assigned to roles that have more rights than the subject should have. A student assistant had the role of a researcher.
- The ad-hoc role $AD - HOC$ is a technical role specific to the object "International". It is not used anywhere else and contains a reference to a subject that does no longer exist in the directory server.
- There exist a number of pseudo-subjects, such as test accounts that allow system administrators to impersonate members of specific roles.

---

[26] subjects * objects * access rights $\approx 4000 \cdot 2 \cdot 6$.

Table 1. Access Rights on Object "Project X" by Role.

Rights $\mathcal{R}$	Role	Subjects in Role
f, m, e, l, r, w	$Role_A$	16
f, m, e, l, r, w	$Role_B$	13
f, m, e, l, r, w	$Role_C$	42
f, m, e, l, r, w	$Role_D$	4
e, l, r	$Role_E$	13

Table 2. Access Rights on Object "International" by Role.

Rights $\mathcal{R}$	Role	Subjects in Role
f, m, e, l, r, w	$Role_A$	16
e, l, r	$Role_F$	193
e, l, r	$Role_G$	150
f, m, e, l, r, w	$Role_H$	64
e, l, r	$Role_I$	3983
e, l, r	$Role_J$	172
e, l, r	$Role_K$	305
e, l, r	$Role_L$	178
e, l, r	$Role_M$	65
m, e, l, r, w	$AD - HOC$	5

- Technical roles (subjects with the same rights) and organizational roles (subjects with the same job position) are mixed arbitrarily.

This list of discrepancies illustrates how error-prone the maintenance of join, move and leave operations is in RBAC. For each operation, all affected role assignments have to be maintained. In $\mathcal{HGAC}$, these operations have to be performed *once* in the organizational model.

Figure 5 shows the hypergraph in $\mathcal{HGAC}$ that is equivalent to the representation in RBAC. For clarity, the different sets of rights $\{e, l, r\}$, $\{f, m, e, l, r, w\}$ and $\{m, e, l, r, w\}$ are represented as one relation-type each in the depiction. The actual hypergraph contains a relation-type per access right.

The key of the representation of an RBAC model in $\mathcal{HGAC}$ is the formulation of the roles as language expressions:

- $Role_A$ represents the IT-administrators of the University: `Admin (University)`
- $Role_B$, $Role_C$ and $Role_D$ are technical roles for members of the Research Department: `Member(Research Department)`
- $Role_E$ are employees of the IT-Infrastructure department: `Member (IT- Infrastructure)`
- $Role_F$ and $Role_G$ encompass lecturers of the university: `Lecturer(*)`.
- $Role_H$ are employees of the Datacenter department: `Member(Datacenter)`.
- $Role_I$ $Role_J$ and $Role_K$ represent different types of students, e.g. external and internal students. As they appear together, they can be represented as `Student(*)`.
- $Role_L$ represents all administrative employees of the university, `*(Administration)`.
- $Role_M$ are research assistants of the University, as can be described as `Research Assistant(University)`.

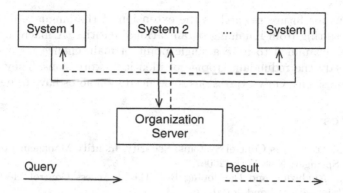

Fig. 6. Organizational Server connected with Application Systems adapted from [20].

- The special role $AD-HOC$ contains all employees of the International Office of the university. The expression *(International Office) describes them.

6 Conclusion

The approaches ACM and RBAC require extensive *maintenance effort* (cf. Eqs. (1)–(3)) to accommodate fluctuating subjects in the company. This effort is erased in $\mathcal{HGAC}$ (cf. Eq. (15)). The language expressions declaring policies/access rights are more stable over time than total enumeration used by ACM and RBAC. Access rights are defined in $\mathcal{HGAC}$ in a descriptive manner. The description can be based on organizational structures (i.a. departments, job functions, supervisor relations) and properties (i.a. name, hiring year, salary). In addition, parameters from application systems or different contexts can be decisive. The contexts are defined by the nature of necessary tasks or of common resources. The organizational model includes the definition of responsible subjects for the administration of the organizational model itself. This decreases the workload of administrators and distributes the work to the subjects that maintain the organizational model.

The subject that joins, moves or leaves the company causes maintenance effort in all application systems respectively applications which is prone to error. As a consequence, the definition of *consistent access rights* that conform to reality is facilitated.

The concept of $\mathcal{HGAC}$ in conjunction with an organizational server solves the afore-mentioned problems (cf. Fig. 6). The systems hand the language expressions (query) to the organizational server holding the organizational model of the company. The expressions are evaluated on the model and the set of authorized subjects are handed back to the system (result). All connected systems are immediately in synchronization with organizational facts if the new organizational conditions are modeled in the organizational server. This makes access rights consistent over various application systems.

The focus for future research is the extension of the language to overcome problems resulting from renaming organizational entities. A macro-like mechanism will be examined to have a single point of maintenance for expressions. Macros remedy the redundant storage of identical expressions. They are references to expressions. Only expressions referenced by macros have to be changed.

References

1. Benantar, M.: Access Control Systems: Security, Identity Management and Trust Models. Springer, New York (2006)
2. Chen, L.: Analyzing and Developing Role-Based Access Control Models. Ph.D. thesis, University of London (2011)
3. Chen, Y., Zhang, L.: Research on role-based dynamic access control. In: Proceedings of the iConference, iConference 2011, pp. 657–660. ACM, New York (2011)
4. Ferraiolo, D., Kuhn, D.R., Chandramouli, R.: Role-Based Access Control. Artech House Computer Security Series. Artech House, Norwood (2003)
5. Ferraiolo, D.F., Barkley, J.F., Kuhn, D.R.: A role-based access control model and reference implementation within a corporate intranet. ACM Trans. Inf. Syst. Secur. **2**, 34–64 (1999)
6. Ferraiolo, D.F., Sandhu, R., Gavrila, S., Kuhn, D.R., Chandramouli, R.: Proposed NIST standard for role-based access control. ACM Trans. Inf. Syst. Secur. **4**, 224–274 (2001)
7. Ferrari, E.: Access Control in Data Management Systems. Synthesis Lectures on Data Management. Morgan & Claypool, San Rafael (2010)
8. Fowler, M.: Domain-Specific Languages. Addison-Wesley Professional, Boston (2010)
9. Gallo, G., Longo, G., Pallottino, S., Nguyen, S.: Directed Hypergraphs and Applications. Discrete Appl. Math. **42**(2–3), 177–201 (1993)
10. Graham, G.S., Denning, P.J.: Protection: principles and practice. In: Proceedings of the May 16–18, Spring Joint Computer Conference, AFIPS 1972 (Spring), pp. 417–429. ACM, New York (1972)
11. Hoffmann, D.W.: Theoretische Informatik, 2nd edn. Carl Hanser, München (2011)
12. Knorr, K.: Dynamic access control through petri net workflows. In: 16th Annual Conference of Computer Security Applications, ACSAC 2000, pp. 159–167, December 2000
13. Krcmar, H.: Informationsmanagement. Springer, Heidelberg (2010)
14. Lawall, A., Reichelt, D., Schaller, T.: Intelligente Verzeichnisdienste. In: Barton, T., Erdlenbruch, B., Herrmann, F., Müller, C. (eds.) Herausforderungen an die Wirtschaftsinformatik: Betriebliche Anwendungssysteme, AKWI, pp. 87–100. News & Media, Berlin (2011)
15. Lawall, A., Reichelt, D., Schaller, T.: Propagation of agents to trusted organizations. In: 2014 IEEE/WIC/ACM International Joint Conferences on Web Intelligence (WI) and Intelligent Agent Technologies (IAT), vol. 3, pp. 433–439, August 2014
16. Lawall, A., Schaller, T., Reichelt, D.: An approach towards subject-oriented access control. In: Stary, C. (ed.) S-BPM ONE – Scientific Research. LNBIP, vol. 104, pp. 33–42. Springer, Heidelberg (2012)
17. Lawall, A., Schaller, T., Reichelt, D.: Integration of dynamic role resolution within the S-BPM approach. In: Fischer, H., Schneeberger, J. (eds.) S-BPM ONE - Running Processes. CCIS, vol. 360, pp. 21–33. Springer, Heidelberg (2013)

18. Lawall, A., Schaller, T., Reichelt, D.: Who does what - comparison of approaches for the definition of agents in workflows. In: 2013 IEEE/WIC/ACM International Joint Conferences on Web Intelligence (WI) and Intelligent Agent Technologies (IAT), vol. 3, pp. 74–77, November 2013
19. Lawall, A., Schaller, T., Reichelt, D.: Cross-organizational and context-sensitive modeling of organizational dependencies in C-ORG. In: Nanopoulos, A., Schmidt, W. (eds.) S-BPM ONE - Scientific Research. LNBIP, vol. 170, pp. 89–109. Springer, Heidelberg (2014)
20. Lawall, A., Schaller, T., Reichelt, D.: Enterprise architecture: a formalism for modeling organizational structures in information systems. In: Barjis, J., Pergl, R. (eds.) EOMAS 2014. LNBIP, vol. 191, pp. 77–95. Springer, Heidelberg (2014)
21. Lawall, A., Schaller, T., Reichelt, D.: Local-global agent failover based on organizational models. In: 2014 IEEE/WIC/ACM International Joint Conferences on Web Intelligence (WI) and Intelligent Agent Technologies (IAT), vol. 3, pp. 420–427, August 2014
22. Lawall, A., Schaller, T., Reichelt, D.: Restricted relations between organizations for cross-organizational processes. In: 2014 IEEE 16th Conference on Business Informatics (CBI), Geneva, pp. 74–80, July 2014
23. Liu, Y.A., Wang, C., Gorbovitski, M., Rothamel, T., Cheng, Y., Zhao, Y., Zhang, J.: Core role-based access control: efficient implementations by transformations. In: Proceedings of the ACM SIGPLAN Symposium on Partial Evaluation and Semantics-Based Program Manipulation, PEPM 2006, pp. 112–120. ACM, New York (2006)
24. Sandhu, R.S.: The typed access matrix model. In: Proceedings of the IEEE Symposium on Security and Privacy, SP 1992, pp. 122–136. IEEE Computer Society, Washington, D.C. (1992)
25. Sandhu, R.S.: Role-based access control. Adv. Comput. 46, 237–286 (1998)
26. Sandhu, R.S., Coyne, E.J., Feinstein, H.L., Youman, C.E.: Role-based access control models. Computer 29(2), 38–47 (1996)
27. Saunders, G., Hitchens, M., Varadharajan, V.: Role-based access control and the access control matrix. SIGOPS Oper. Syst. Rev. 35(4), 6–20 (2001)
28. Seufert, S.E.: Die Zugriffskontrolle. Ph.D. thesis, Bamberg, Univ., Diss., (2002)
29. Vahs, D.: Organisation: Einführung in die Organisationstheorie und -praxis. Schäffer-Poeschel (2007)
30. Williamson, G., Sharoni, I., Yip, D., Spaulding, K.E.: Identity Management: A Primer. MC Press Series. MC Press Online, Big Sandy (2009)

NoSQL Approach to Data Storing in Configurable Information Systems

Sergey Kucherov[✉], Yuri Rogozov, and Alexander Sviridov

Department of System Analysis and Telecommunications,
Southern Federal University, 44 Nekrasovsky St., Taganrog, Russian Federation
{skucherov, yrogozov, asviridov}@sfedu.ru

Abstract. The problem of data storing in configurable information systems is solved today by introducing redundancy in the storage structure and interaction logic between information system and storage. Causes which are forcing developers of configurable information systems to introduce such a redundancy are rooted in diverse ways to present system parts and the absence of a single coherent system model. To avoid redundancy and to create new generation of configurable information systems an unified abstraction. Studies have shown that the most appropriate abstraction to a subject area as well as to information systems is the action abstraction. An information system inherently is nothing but a reflection of user actions with the help of computer technology.

The paper proposes a new approach to data warehousing in configurable information systems which is based on the action abstraction and NoSQL technology. Warehouse model and basic principles of its functioning are presented.

Keywords: Data storage · NoSQL · Configurable information system · GlobalsDB

1 Introduction

The transition to a new generation of information systems (IS) [1] - configurable IS (CoIS) [2], differ both in architecture [3] and the principles of operation [4], causes the appearance of another data storing principles and ways of its implementation [5].

Setting aside technical and methodological features of development processes an IS can be considered as a reflection of user actions performed using other means, but with the same result [6]. We considered classical approaches to a design of flexible software systems [7], and formulated the following abstract scheme of an IS design process from the "main abstraction" [8] viewpoint (Fig. 1).

Initial state of a domain is a performing certain actions by user. First step in transition to an information system is describing user actions in the form of business process model that is fixing his actions. Depending on design concepts representation of business processes can vary, but in any case there is the first step to avoiding the usual to an user "action" abstraction. Next step is representation of a business process as a set of abstractions describing an information system, which is capable to performing the process. Usually, the set abstractions at the modelling stage is the structural

M. Helfert et al. (Eds.): DATA 2015, CCIS 584, pp. 120–134, 2016.
DOI: 10.1007/978-3-319-30162-4_8

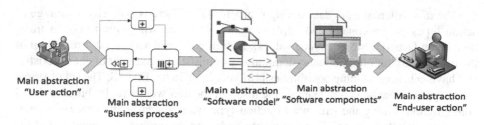

Fig. 1. Abstract scheme of the classical IS design process.

(IDEF, DFD) or object (UML) model. At the design and implementation stage abstractions are interface, business logic and data, which is absolutely not comparable with user's "action" abstraction. Despite the fact that as a result we change user action (performed manually) to end-user action (performed by an information system), the form of presentation varies significantly. This variety creates a lot of problems, and the main problem is misunderstanding between an user and a developer. The reasons are lies in application of classical analysis methods to nonclassical problem for them [9].

In our opinion, the only abstraction that can represent IS in adequate to a domain and to an user form is the base abstraction for action synthesis (BAfAS) [10]. BAfAS is the fixed set of characteristics with variable content [11]. Changing content determines the assignment of actions, while the fixed structure of an action representation (characteristics: element, function, tool, result) allows to use one abstraction throughout the process of creating an IS - from a study of subject area, prior to an implementation and an operation of the system [12]. This, in turn, will require different methods of system analysis and synthesis based on the "action" abstraction. The paper introduces the concept of configurable information system's storage based on the "action" abstraction and NoSQL technologies.

Configurable information system should primarily provide the flexibility to customizing, extending and changing functionality with minimal effort [13]. In this context, there is necessity to shift from the traditional IS partitioning to "interface", "business logic" and "data" to a common base abstraction [14]. Base abstraction should be the unified way of an IS representation. For many years there was the problem of misunderstanding between developer and end-user [15], caused by substitution of concepts in the implementation of the system: the end-user requires automation of specific actions or his entire activity, while the developer operates mentioned above aspects – "interface", "business logic", "data" [16].

BAfAS allows to simulate the user activity, but not parts of an information system, reproducing the user's actions (as it does a lot of existing methodologies). This is the principally new viewpoint on information systems [6]. To describe an end-user's activity BAfAS uses a simple set of characteristics - elements, functions, tools and results. Regardless of the detalization level this set stays fixed and represents the shape. Filling the BAfAS shape with content (specification of elements, functions, etc.) is a process of construction an action. Actions could form the activity by general characteristics, by binding through "result = element" or by nesting in each other.

The task different from data storing is arising [17]. There is necessity to storage of action types, filling content and implementation of which will allow to obtain traditional data facts (i.e. stored named values). Action types and actions (as a reflection of the user work with information system), always consist of a fixed set of characteristics and have enhanced binding possibilities [18] laid down in the BAfAS. The set of stored action types and actions is not simply the result of the user work with IS, but also the IS configuration. Filling the structure of action type with content is the process of configuring an IS. Performing of action and remembering this process is a reflection of the user work with IS. The configurable information system's storage should have the next number of new properties:

- Shifting the focus in storage: key aspect is action type (reflects configuration of the IS) and the action itself (lead to obtain the fact), not some fact taken separately (result of action);
- Expansion of the connection's significance between stored objects. In the classical approach the relationship between instances of stored data objects means their mutual affiliation. In the case of storing actions connection determines the sequence of steps in automated end-user activity.

The possibility of using widespread data storage technologies for obtaining of the above properties requires analysis. In the paper we represent CoIS storage concept that are based on the BAfAS and NoSQL-technologies. The action storing model and requirements to CoIS storage is formulated in the second section. The third section presents the results of data storage technologies analysis in case of their application to solve described problem. The fourth section describes a conceptual model of CoIS storage.

2   The Model of Action Representation in Configurable Information Systems

The data model (the most close in meaning term) includes three aspects [19] - data structure, data manipulation rules and tools to ensure integrity. At the stage of formulating action storage conception in CoIS we will focus on the structural aspect. The use of classical approaches for creating configurable information systems invariably leads to complication turnkey solutions from a technical point of view [20]. Despite the apparent simplicity of operation configurable systems, their support without the involvement of technical experts is difficult.

One reason for the complexity of created solutions is no single view (abstraction) on an IS. Consequence of this is multiple interfaces to make an adequate state of system components (e.g., ORM-layer, resulting a relational database adequate to the object model). The system should be presented in a unified manner to resolve this issue. This manner could be a base abstraction, which is a reflection of user's actions. Base abstraction should be used to present both logic of interaction with user, stored action results and configuration of the system itself. The actions that can be performed, but does not include values of characteristics represent the system configuration. Actions that have been performed with certain values of characteristics - represent the result of user work with the information system [6].

Consequently, the model of action representation in configurable information system, being adequate to unified BAfAS model, should use the action as a basic storage unit. From the structural aspect in the model of action representation can be distinguished: action type, action (performed), constant.

Action type is a reflection of the user's work unit, expressed by:

- used elements:
- applied functions;
- tool that regulates rules of applying functions to elements;
- the result that can be obtained by performing the action.

Result corresponds to the purpose of action implementation. It should be noted that the action type has only specified the expected outcome in the form of performing goals, while the value of the result appears after the action implementation. Action type is the basic unit of storage and representation in configurable information system. An example of the user's action type can be "calculation of the monthly wage of an employee" (Fig. 2).

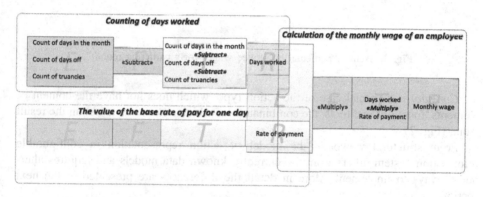

Fig. 2. Action type "calculation of the monthly wage of an employee".

As elements the action type uses information on number of days worked (which in turn is the result of Action type "counting the number of days worked") and information about the cost of one working day (which is the constant - action type with non-specified elements, functions, and tools, but with specified result value). As functions the action type uses the mathematical operator "multiplication". The tool in this case is a multiplication rule, included in one of the system modules.

In case of representing a system as actions erased a clear separation on the "data", "business logic" and "interface". The IS model becomes more adequate to picture of the real world. Described action type can be repeatedly reproduced to calculation of the monthly wage. In such context, the action type is much like the concept of the function in a programming language, but unlike function the action is not a fragment of code which access data and GUI across multiple interfaces. Action is independent stored element, which can be configured and used as a source of values.

Action is a result of performing the action type at any point in time, containing specific values of all characteristics, including the value of result (Fig. 3). In the context of above described example, the action can be a particular employee wages for a particular month. BAfAS will allow to store not only the wage value, but also its production process. This largely enhances the warehousing possibilities in terms of maintaining the integrity and historicity changes.

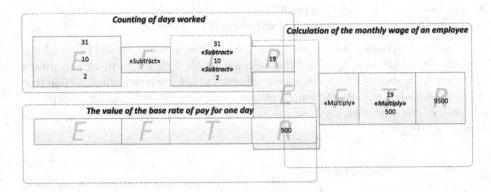

Fig. 3. Action "particular employee wages for a particular month".

Constant is a special case of the action type, which does not have the content of elements, functions and tools, the combination of which allowed us to obtain the result value (Fig. 2).

From structural viewpoint the model of action representation in configurable information system differs from the currently known data models and requires alternative ways to implement. More in detail the differences are presented in the next section.

3   Analysis of the Applicability for Existing Storage Technologies

In order to evaluate the applicability of existing storage technologies for configurable information system we denote key requirements dictated by the model of action representation [21]:

- Fixed structure of action representation. Regardless of the action content, it must always be presented in four characteristics - elements, functions, tools and results.
- Any of action characteristics actions may have no content or the content may be presented by a plurality of elements with different types;

- As the value type for each of characteristics (except the result) may be action results or a whole action. Herewith in particular action can be used as single types, and various combinations thereof.

The basic storage unit should meet these requirements, because in the other case it will again cause a problem complicating the system [22], and may also exhibit negative performance [23]. From the many created for today data storage technologies clearly the declared basic storage unit only have two - relational and object-oriented. In the first case the basic storage unit is a relation that allows to group and record the individual facts. In the second case, the basic storage unit is a class that contains a set of attributes and operations. Results of analysis mentioned above data storage technologies from the viewpoint of BAfAS implementation are shown in Table 1.

Table 1. Results of analysis.

Requirement	Relational technology	Object-oriented technology
Fixed structure	Unavailable. The structure of relationship is determined at the time of its setting on domains, the number of which can be arbitrary	Available. Class is always represented by the name, attributes and methods
Multiple values of characteristics	Unavailable. Relation instance contains one value of each domain	Partially available. Class instance is characterized by a specific attribute values. The problem can be solved by complex data types
The multiplicity of values types for a single characteristic	Unavailable. Each domain is a separate data type	Partially available. Class attribute has a specific value type. The problem can be solved by complex data types

None of the submitted data storage technology is not able to meet the requirements. This is a consequence of the narrow focus and it affects the final form of the base storage unit [24].

To meet the model requirements configurable information system's storage technology can be obtained in one of two ways:

- development of its own storage technology which has adequate representation in the form of action;
- use of existing storage technologies, allowing to independently define the base storage unit structure.

First method is more efficient in terms of the final result, second method has fewer risks and deadlines. As part of the study is expected to first carry out optimization and testing of the model on existing technologies, and then go on to create their own technology.

For today there is a whole class of models and data storage technologies, combined by the term NoSQL [25]. These solutions include next types: family of columns, key-value, document-oriented etc. Because information systems are focused on factual information storage, document-oriented ways of presenting information is not applicable, in addition, they require additional processing after extraction of information (document) from the storage. Actions have a clear structure and themselves, in turn, form a complex structure - user activity. Therefore, the key-value technology is not applicable too, since storage is not natively focused on associated data objects. Mentioned requirements are met by column family-based types of NoSQL storage.

NoSQL storages oriented to work with families of columns have different ways of presenting data [26]. Of greatest interest are so-called schema-free systems [27, 28], wherein the base storage unit has no deterministic structure of representation. The unit structure can be initially set in the adequate to configurable information system form. Such systems use basically principles laid down in the hierarchical and network data models - each data object is a sparse hierarchical tree of elements. Next, consider the conceptual model of the configurable information system's storage based on NoSQL-technology.

4   Conceptual Model of the Configurable Information System's Storage

Both action types and actions must meet BAfAS requirements. In this connection, convenient to introduce the storage concept as a multi-layered storing system of actions (Fig. 4). The base of the storage (also the basic storage unit and the first layer) is an BAfAS, consisting of four characteristics - elements, functions, tools and results.

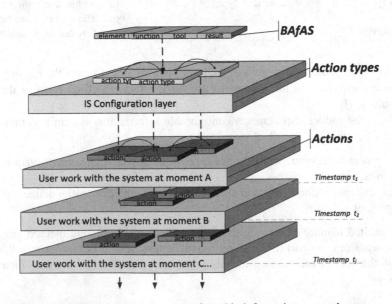

Fig. 4. Conceptual model of the configurable information system's storage.

The next layer is the system configuration, expressed by action types. Each action type is the result of applying the basic abstraction to non-automated user activity. Action type contains elements, functions and tools, which implementation will lead to obtain the target result. Every action type can be repeatedly reproduced thereby will be memorized actions reflecting the user work with the system.

The system configuration representing by connected action types (base layer) are created once at the configuring stage. During the IS operation we are building up the layers reflecting results of a different action types in variable contexts (secondary layer). Each secondary layer is a result of action types implementation fixed at a certain point in time. Horizontal upbuilding of the base layer occurs during system configuration by adding new action type. Vertical upbuilding occurs in IS operation (in new times we perform action types from the base layer). Multi-layered character of the storage is a consequence of using the unified abstraction in the process of creating and operating the system. Next consider models of action type, action and constant taking into account the NoSQL technology.

Figure 5 shows the model of BAfAS representation in NoSQL configurable information system's storage. Action type is formed through the common pattern (base abstraction for action synthesis), which includes the name of the action, and a list of four characteristics with their constituent components. Result in the description of action type is degenerate, since the action is described, but not yet implemented. Due to schema-free technology the number of components and their types can be determined dynamically.

Action, in contrast to an action type, has values of characteristics. Each value is accompanied by a time-stamped obtaining or assignment. Action also store the result of

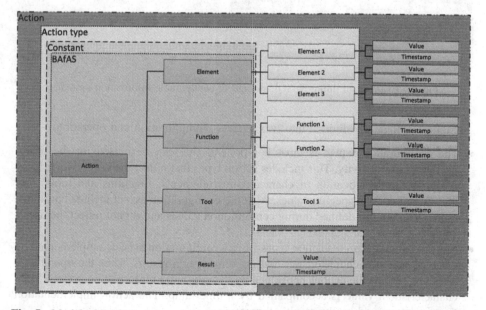

Fig. 5. Model of BAfAS representation in NoSQL configurable information system's storage.

its implementation, and the timestamp of result value is considered as the moment of action implementation. Constants are a particular example of an action type, because they are recorded at the configuration stage. They have degenerated characteristics of elements, tools and functions and contain only the result with the timestamp.

As we seen in Fig. 5, each base storage unit is a named tree that describes the user action. Inside a node may be hierarchy describing the embedding of actions at each other and allowing to describe complex user activities.

Described above components are basic components of the conceptual model of the configurable information system's storage. Storage itself is done at the expense of hierarchical trees constructing. Next, the example from of the first part shows the general concept of the storage (Fig. 6).

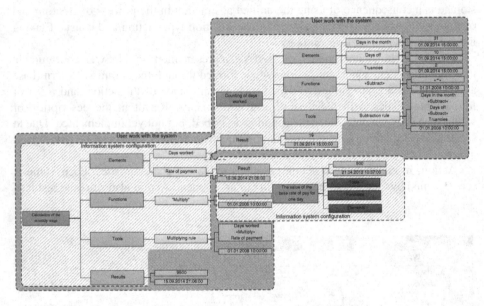

Fig. 6. Example of filling NoSQL storage of configurable information system.

Storage is logically divided into two aspects - configuration and operation.

- The configuration aspect is those components that can be reused for the reproduction of user activity. This includes action types that reflect the user's work. They comprise the specification of characteristics (elements, functions and tools), but they do not contain specific values. Also the configuration aspect include constants, as they contain pre-defined during configuration values and do not reflect the user's work.
- The operational aspect is action types from a configuration aspect, supplemented by value of specified characteristics and implementation results. Thus the operational aspect represents the actions performed by the user within information system. Each characteristic's value is accompanied by a timestamp. The action implementation time is determined by the result's timestamp.

Drawing an analogy with object-oriented and relational technologies it can be said that the action type (by destination) is like a class (entity), and the action is like an object (entity instance). However, unlike entities and classes the action storage model has significantly greater opportunities:

- it is more dynamic;
- it has more semantics than the current data models;
- it is more adequate to picture of the real world.

Main ideological difference of the proposed storage is that it is not based on a set of separated actions, but on BAfAS, which allows to receive various types of user actions for further reproduction.

The proposed storage allows the use of different types of characteristic's values, thereby building a sequence of action types, the implementation of which leads to a result (data fact). By using BAfAS we met requirements of the action representation model in configurable information system, and also the conditions for its implementation without additional funds to bring an adequate form (interfaces). The possibility of "growth" at the expense of schema-free allows unlimited detalization of actions. It also allow to describe complex activities.

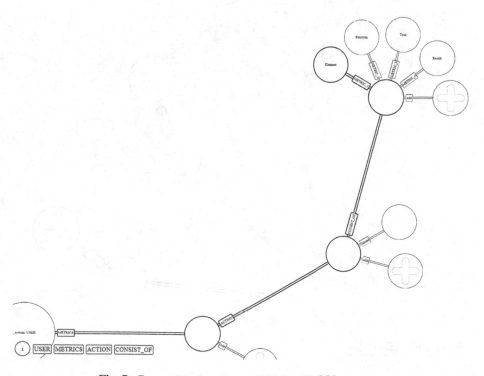

Fig. 7. Pattern "Action abstraction" in a NoSQL storage.

5 NoSQL Patterns for Warehouse Implementation

Implementation of storage is carried out on the base of GlobalsDB DBMS [27, 28]. A key feature of this solution is the full supporting of schema-free approach. At the heart of the repository are permanently stored global variables – globals.

The approach to implementation of described storage model is based on creation of a special data structure. This structure is a multidimensional sparse array. Structural elements of the array are the metrics chosen to describe the subject domain. In our case, this metrics - user action's characteristics.

In accordance with the ideology of action abstraction, it is subdivided into elements, functions, tools and result. Action abstraction (Fig. 7) is a basic pattern for the model of user actions delegated for an information system. Action abstraction can take next forms: action type, action, and constant (as described in Sect. 2).

Connection of actions are carried out by a special system "IS_A_RESULT" global. The value of "IS_A_RESULT" global is a link to other global containing desired value. There is an important feature in representing a system by set of actions - any characteristic of action abstraction may be the result of another action. This significantly extends classical representation of data processing, which applies only one connection type – "result-element".

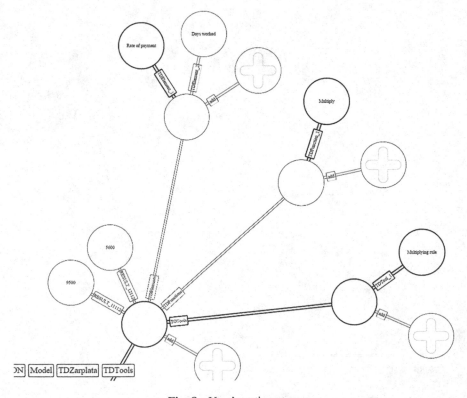

Fig. 8. User's actions tree.

Creation of user actions model is a process of sequential building the tree by instances of the pattern "action abstraction". So, to set described payroll model was created next actions tree (Fig. 8).

This tree contains both the specification and user actions in the form of their types, and results of actions. The system configuration layer in the form of the action types is described by using globals with prefixes named TD*. Results of system working process are store in globals prefixed name RESULT_ *, where "*" is an absolute time, set at the time of result obtaining. This set of globals is a preliminary stage of operation of the repository.

6 Conclusions

For today the creation of flexible information system is the process of introducing redundancy in the storage structure and interaction logic between information system and storage. Such redundancy is a result of application of classical analysis methods to nonclassical problem for them. It creates a lot of difficulties - from the loss of productivity due to overhead costs to the reduction of fault-tolerance.

The paper proposed the basis for storing data in the cardinally new way – by using abstraction of action synthesis (BAfAS) because of an information system inherently is nothing but a reflection of user actions with the help of computer technology. Accordingly, a model of information system is a set of complexly linked actions. And the process of its creation - is a realization of user action properties with the help of computer technology. Similar to the information system model, a reflection of user's works with the system are also a sequence of committed actions.

Since an information system model and the user's actions are represented by a single abstraction and have dynamic in nature, they are needed to be kept in a uniform way. The aim of storing in this case is different from the existing today. We don't need to store individual values. Data warehouse of configurable information system should allow to memorize a set of complexly linked actions with the possibility of its multiple executions.

Representation of configurable information system, storing of its configuration and its operating results as a set of actions will eliminate currently existing problems, such as: growth of the technical complexity with enhancing flexibility, misunderstanding between end-user and developer, extraction requirements, etc.

BAfAS allows to bring the storage to a new level - data semantics, their occurrence are clearly documented in a readable form for an user. This in turn has a positive effect on both the modeling process, and the process of operation and improvement of the system by end-user.

The model of action representation in configurable information system and storage model proposed in this paper have the following advantages:

- Storing data in an adequate form to information system model;
- Eliminate the need for additional mechanisms of transformation and interfaces between the storage and information system.

Described NoSQL storage prototype shows, that idea of whole system represen-tation by actions has real possibilities of implementing. But anyway there is a signif-icant list of challenges to be solved for storage creation. This list consist of next major items:

- the creation of set of stored procedures to implementation of manipulation rules for user actions and its structure;
- developing of "action integrity" similarly to data integrity. Cause there is more complex dependencies between actions;
- developing of case tools for modeling the whole information system in terms of actions.

There is one problem in flexible data storing that was already solved by NoSQL schema-free technologies. It allows to realize high-performance solutions devoid of drawbacks structure-independent databases [17, 21].

Acknowledgements. The reported study was partially supported by RFBR, research project No. 15-07-04033.

References

1. Rogozov, Y., Degtyarev, A.: The basic foundation of software framework for configuration underwater acoustic information systems with dynamic structure. In: Information and Communication Technology for Education (ICTE-2013), vol. 58, pp. 181–189. WIT Press, Southampton (2014)
2. Garcia, J., Goldszmidt, G.: Building SOA composite business services (2007)
3. http://www.ibm.com/developerworks/webservices/library/ws-soa-composite/
4. Rogozov, Y., Degtyarev, A.: Evaluating the effectiveness of building a software sonar information systems using configurable software framework. Engineering Gazette Don (2013). http://www.ivdon.ru/magazine/archive/n4y2013/1877
5. Sviridov, A.: Configuration of information systems in terms of control systems. News of Southern Federal University, Technical Sciences, no. 6, pp. 168–173 (2014)
6. Kucherov, S., Rogozov, Y., Sviridov, A.: The method of configuring dynamic databases. WIT Trans. Inf. Commun. Technol. **58**, 163–173 (2014). WIT Press, Boston
7. Rogozov, Y.I.: Approach to the definition of a meta-system as system. Proc. ISA RAS – 2013 **63**(4), 92–110 (2013)
8. Rossi, C., Guevara, A., Enciso, M.: A tool for user-guided database application development. Automatic design of XML models using CBD. In: Proceedings of the 5th International Conference on Software and Data Technologies, ICSOFT 2010, vol. 2, pp. 195–200 (2010)
9. Rogozov, Y., Sviridov, A., Belikov, A.: Approach to CASE-tool building for configurable information system development. In: Information and Communication Technology for Education (ICTE-2013), vol. 58, pp. 173–181. WIT Press, Southampton (2014)
10. Lazarev, V., Rogozov, Y., Sviridov, A.: Methodological approach as the successor of object approach in development of information systems. J. Inf. Commun. **2**, 85–88 (2014)
11. Rogozov, Y.: Methodology of creation of subject-oriented systems. J. Inf. Commun. **2**, 6–10 (2014)

12. Rogozov, Y., Sviridov, A.: Approach to the construction of information systems based on the methodological approach. In: Proceedings of 2nd International Conference Innovative Technologies and Didactics in Teaching, vol. 1, S.3–9. Publishing House SFU, Taganrog (2014)

13. Rogozov, Y., Sviridov, A.: Methodological concept of building information systems. J. Inf. Commun. 2, 11–14 (2014). Rogozov, Y., Degtyarev, A.: The method of configuring the functionality of software sonar information systems. News of Southern Federal University, Technical Sciences, No. 1, pp. 13–18 (2014b)

14. Fischer, G., Giaccardi, E.: Meta design: a framework for the future of end user development. In: Lieberman, H., Paterno, F., Wulf, V. (eds.) End User Development. HCIS, vol. 9, pp. 427–457. Springer, Heidelberg (2006)

15. Rogozov, Y.: The general approach to the organization of certain system of concepts based on the principle of generating knowledge. In: Proceedings of 12th All-Russian Conference on Control VSPU-214, Moscow, pp. 7822–7833, 16–19 June 2014. ISBN: 978-5-9145-151-5

16. Kucherov, S., Lipko, J., Schevchenko, O.: The integrated life cycle model of configurable information system. In: Proceedings of IEEE 8th International Conference on Application of Information and Communication Technologies - AICT2014, IEEE Catalog Number CFP1456H-PRT, pp. 182–186 (2014). ISBN: 978-1-4799-4120-9 2

17. Kucherov, S.: User-configurable information systems as a means to overcome the semantic gap. J. Inf. Commun. 5, 135–137 (2013)

18. Kucherov, S., Samoylov, A., Grishchenko, A.: Flexible database for configurable information systems. In: Proceedings of IEEE 8th International Conference on Application of Information and Communication Technologies – AICT 2014, IEEE Catalog Number CFP1456H-PRT, pp. 374–377 (2014). ISBN: 978-1-4799-4120-9 2

19. Kucherov, S., Sviridov, A., Belousova, S.: The formal model of structure-independent databases. In: Proceedings of 3rd International Conference on Data Management Technologies and Applications, pp. 146–152. Scitepress - Science and Technology Publications, Vienna (2014). ISBN: 978-989-758-035-2

20. Codd, E.F.: Data models in database management. In: Brodie, M.L., Zilles, S.N. (eds). Proceedings of Workshop in Data Abstraction, Databases, and Conceptual Modelling, Pingree Park, Colo (June 1980): ACM SIGART Newsletter No. 74 (January 1981); ACM SIGMOD Record 11(2), February 1981; ACM SIGPLAN Notices 16(1), January 1981

21. Shaw, M., Garian, D.: Software Architecture: Perspectives on an Emerging Discipline. Prentice Hall, Englewood Cliffs (1996). ISBN: 0-13-182957-2

22. Rogozov Y., Kucherov S., Komendantov, K.: The methodological approach to creation of user data structures in user-configurable information systems Innovative technologies and Didactics in Teaching: collected papers, pp. 171–184. MVB Marketing- und Verlagservice des Buchhandels GmbH, Berlin (2014)

23. Rogozov, Y., Sviridov, A., Grishchenko, A.: The method of data manipulation operations representation as a structure in structure-independent databases oriented on configurable information system development. WIT Trans. Inf. Commun. Technol. 58, 189–197 (2014). WIT Press, Boston

24. Grishchenko, A., Rogozov Y.: Criterial rating of the data manipulation methods for configurable information system development. Innovative technologies and Didactics in Teaching: collected papers, pp. 46–52. MVB Marketing- und Verlagservice des Buchhandels GmbH, Berlin (2014)

25. Kucherov, S., Rogozov, Y., Borisova, E.: Structure-independent databases modelling. In: Proceedings of IEEE 8th International Conference on Application of Information and Communication Technologies - AICT 2014, IEEE Catalog Number CFP1456H-PRT, pp. 192–196 (2014). ISBN: 978-1-4799-4120-9 2
26. Sadalage, P.J., Fowler, M.: NoSQL Distilled: A Brief Guide to the Emerging World of Polyglot Persistence (2012). ISBN: 13 978-0321826626
27. McCreary, D., Kelly, A.: Making Sense of NoSQL: A Guide for Managers and the Rest of Us, pp. 1–312. Manning Publications, Greenwich (2013). ISBN: 978-1-61729-107-4
28. Tweed, R., George, J.: A Universal NoSQL Engine, Using a Tried and Tested Technology (2010). http://www.mgateway.com/docs/universalNoSQL.pdf

Needles in the Haystack — Tackling Bit Flips in Lightweight Compressed Data

Till Kolditz[1], Dirk Habich[1(✉)], Dmitrii Kuvaiskii[2], Wolfgang Lehner[1], and Christof Fetzer[2]

[1] Technische Universität Dresden, Database Systems Group,
01062 Dresden, Germany
{Till.Kolditz,Dirk.Habich,Wolfgang.Lehner}@tu-dresden.de
[2] Technische Universität Dresden, Systems Engineering Group,
01062 Dresden, Germany
{Dmitrii.Kuvaiskii,Christof.Fetzer}@tu-dresden.de

Abstract. Modern database systems are very often in the position to store their entire data in main memory. Aside from increased main emory capacities, a further driver for in-memory database system has been the shift to a column-oriented storage format in combination with lightweight data compression techniques. Using both mentioned software concepts, large datasets can be held and efficiently processed in main memory with a low memory footprint. Unfortunately, hardware becomes more and more vulnerable to random faults, so that e.g., the probability rate for bit flips in main memory increases, and this rate is likely to escalate in future dynamic random-access memory (DRAM) modules. Since the data is highly compressed by the lightweight compression algorithms, multi bit flips will have an extreme impact on the reliability of database systems. To tackle this reliability issue, we introduce our research on error resilient lightweight data compression algorithms in this paper. Of course, our software approach lacks the efficiency of hardware realization, but its flexibility and adaptability will play a more important role regarding differing error rates, e.g. due to hardware aging effects and aggressive processor voltage and frequency scaling. Arithmetic AN encoding is one family of codes which is an interesting candidate for effective software-based error detection. We present results of our research showing tradeoffs between compressibility and resiliency characteristics of data. We show that particular choices of the AN-code parameter lead to a moderate loss of performance. We provide evaluation for two proposed techniques, namely AN-encoded Null Suppression and AN-encoded Run Length Encoding.

1 Introduction

Data management is a core service for every business or scientific application in today's data-driven world. The data life cycle comprises different phases starting from understanding external data sources and integrating data into a common database schema. The life cycle continues with an exploitation phase

© Springer International Publishing Switzerland 2016
M. Helfert et al. (Eds.): DATA 2015, CCIS 584, pp. 135–153, 2016.
DOI: 10.1007/978-3-319-30162-4_9

by answering queries against a potentially very large database and closes with archiving activities to store data with respect to legal requirements and cost efficiency. While understanding the data and creating a common database schema is a challenging task from a modeling perspective, efficiently and flexibly storing and processing large datasets is the core requirement from a system architectural perspective [17, 28].

With an ever increasing amount of data in almost all application domains, the storage requirements for database systems grows quickly. In the same way, the pressure to achieve the required processing performance increases, too. To tackle both aspects in a consistent uniform way, data compression as software concepts plays an important role. On the one hand, data compression drastically reduces storage requirements. On the other hand, compression also is the cornerstone of an efficient processing capability by enabling "in-memory" technologies. As shown in different papers, the performance gain of in-memory data processing for database systems is massive because the operations benefit from its higher bandwidth and lower latency [1, 6, 14, 18].

Aside from the developments in the data compression domain, the hardware sector has seen important developments, too. Servers with terabytes of main memory are available for a reasonable price, so that the entire data pool in a compressed form can be kept and processed completely in main memory. In order to increase main memory density and to put more functionality into integrated circuits (ICs), transistor feature sizes are decreased more and more. On the one hand, this leads to performance improvements in each hardware generation. On the other hand, ICs become more and more vulnerable to external influences like cosmic rays, electromagnetic radiation, low voltages, and heat dissipation. Data Centers already face crucial error rates in dynamic random-access memory (DRAM) [13, 25] including multi bit flips which cannot be handled by typical SECDED[1] ECC DRAM anymore. These error rates are likely to increase in the future, and will become a major challenge for in-memory database systems. We argue that ECC DRAM alone is not the silver bullet, because the employed codes are statically integrated into hardware with fixed parameters.

Generally, the field of error correcting codes as well as the field of data compression techniques are well-established. In order to tackle the above mentioned resiliency challenge for in-memory database system, we propose to tightly combine existing techniques from the corresponding files in an appropriate way: *resiliency-aware data compression techniques*. Of course, our software approach lacks the efficiency of resiliency-aware hardware realization like ECC DRAM, but its flexibility and adaptability will play a more important role regarding differing error rates, e.g., due to hardware aging effects and aggressive processor voltage and frequency scaling. Our main idea is to minimize the amount of useful information (bits)—using data compression—which are then enriched by redundant information (bits) to protect against bit flips. As a side constraint, our resiliency-aware compressed data should allow to directly work on that data representation without explicitly decompressing and re-encoding the data. This

---

[1] Single-error correcting and double-error detecting.

constraint is important to achieve query processing performance expected by in-memory database systems. Furthermore, error detection should be possible in an online fashion, so that wrong results can be excluded to a well-defined degree.

Our Contribution. To show the potential and challenges in this research direction, we present our first research results in this paper. From the field of error correcting codes, we have chosen the family of arithmetic AN codes as a very promising alternative (or complementary) to ECC DRAM, since its very nature allows to do arithmetic operations – including comparisons – without the need of decoding. Consequently, arithmetic AN codes are suitable for both transactional and analytical database workloads. From the data compression domain, we decided to use two heavily-used lightweight techniques: Null Suppression [1, 22], and Run Length Compression [1]. In detail, our contributions are as follows:

1. We show how to tightly combine arithmetic AN encoding with Null Suppression [1, 22], and Run Length Compression [1] as concrete examples for our *resiliency-aware data compression techniques* or *AN-encoded lightweight data compression techniques*. As we are going to show, the combination approach differs and depends on various factors.
2. We introduce our *AN-encoded data compression techniques* in our data compression modularization concept. Generally, our modularization concept offers an efficient and an easy-to-use way to describe, to compare, and to adapt (AN-encoded) lightweight data compression techniques.
3. We provide an analysis of how the basic parameter A of AN encoding can be chosen to detect various amounts of bit flips.
4. We show that there are "good" As with very low performance penalties to enable online error detection for compressed data.
5. We provide a performance evaluation for the "good" As for both AN-encoded Null Suppression and AN-encoded Run Length Compression.

Outline. The remainder of this paper is structured as follows: In Sect. 2, we present related work with a brief overview of existing lightweight compression techniques and give a detailed insight into AN encoding in Sect. 3. Then, we present how to integrate AN encoding with two compression schemes in Sect. 4. Next, we present our evaluation in Sect. 5, where we discuss "good" As and provide throughput comparisons for one of the "good" As. Finally, we conclude the paper in Sect. 6.

2   Related Work

Before we present our novel approach of resiliency-aware data compression techniques, we briefly review related work on (1) lightweight data compression techniques frequently used in-memory database system in Sect. 2.1 and (2) generic and database-specific resilience mechanism in Sect. 2.2 and 2.3.

2.1 Lightweight Data Compression Techniques

In the area of conventional data compression a multitude of approaches exists. Classic compression techniques like arithmetic coding [31], Huffman [12], or Lempel-Ziv [32] achieve high compression rates, but the computational effort is high. Therefore, those techniques are usually denoted as heavyweight. Especially for in-memory database systems, a variety of lightweight compression algorithms has been developed. These achieve good compression rates similar to heavyweight methods by utilizing context knowledge, but they require much faster compression and decompression.

The main archetypes or classes of lightweight compression techniques are dictionary compression (DICT) [2,5,16], delta coding (DELTA) [18,22], frame-of-reference (FOR) [7,33], Null Suppression (NS) [1,21,22,27], and Run-Length Encoding (RLE) [3,22]. DICT replaces each value by its unique key. DELTA and FOR represent each value as the difference to its predecessor or a certain reference value, respectively. These three well-known techniques try to represent the original data as a sequence of small integers, which is then suited for actual compression using a scheme from the family of NS. NS is the most well-studied kind of lightweight compression. Its basic idea is the omission of leading zeros in small integers. Finally, RLE tackles uninterrupted sequences of occurrences of the same value, so-called runs. In its compressed format, each run is represented by its value and length, i.e., by two uncompressed integers. Therefore, the compressed data is a sequence of such pairs.

2.2 Generic Resilience Mechanisms

Increased memory density, decreased transistor feature sizes and more are major drivers in the area of hardware development. On the one hand, this leads to performance improvements in each hardware generation. On the other hand, the hardware becomes more and more vulnerable to external influences. As several researches have already stated, especially main memory becomes a severe cause for hardware based failures [13,15,19,25,26]. These errors can be classified into *static or hard errors* as permanently corrupted bits and *dynamic or soft errors* as transiently corrupted bits. In particular, dynamic errors are produced, e.g., by cosmic rays, electromagnetic radiation, low voltage and increased heat dissipation.

While dynamic error rates are still quite low, it is predicted that they increase substantially in the near future [13]. Moreover, dynamic errors already have a significant impact on large-scale applications on massive data sets. The field of fault tolerance against dynamic memory errors is not new and several techniques are well-known. A generally applicable approach is executing the same computation multiple times. In this case, any dynamic error can be detected by comparing the final results – except when both results have the very same error which is just assumed to not happen. The most well-known technique in this class is Triple Modular Redundancy. Error detection and error correction codes represent a second class. In this case, the coding schemes introduce redundancy

to the data [20]. Regarding DRAM bit flips, the most commonly used approach is hardware-based (72,64)-Hamming ECC [20], which realizes single-error
correction and double-error detection (SECDED). Many other general coding
algorithms are available, whereas the enhanced coding schemes are more robust,
however their coding results in higher memory overhead and higher computational costs. Generally, the major problem of ensuring a low dynamic error probability by employing generally applicable techniques is dramatically increasing
costs for memory and computational power.

2.3 Database-Specific Resilience Mechanism

In the past, several techniques have been presented to deal with certain error
classes. To our best knowledge, no research was done in the field of databases to
protect in-memory data against arbitrary bit flips, except our own investigations
on error detecting B-Trees [15]. We presented software based adaptations for
B-Trees, a widely used database index structure, to cope with increasing bit
flip rates in main memory. We showed that pointer sanity checks, parity bits
and checksums can deliver comparable or better error detection on commodity
hardware compared to ECC hardware, since they are able to detect more than
2 bit flips in 8-byte words. Furthermore, we showed that checksums are able to
detect more bit flips and provide higher reliability which is highly desired for
database systems.

  In the field of databases, other relevant related work mainly concentrates on
handling errors during I/O operations, or regards situations where the system
inadvertently writes to wrong memory regions, e.g. due to software bugs like
buffer overflows or broken pointers. For instance, [8,9] harden also the well-known
B-Tree and variants against errors during I/O operations or against certain other
tree corruptions. On the one hand these techniques are offline methods, which
means they are periodically executed. On the other hand, they may be very
heavy-weight, especially when comparing entries between several indices, and
may not be suited for online error detection. Furthermore, bit flips may lead
to false positives and false negatives when querying such trees between these
maintenance checks.

  Sullivan et al. [29] deal with corruptions due to arbitrary writes by employing
hardware memory protection for individual pages. Memory pages are protected
using hardware directives and the protection is removed only when accessing the
pages through a special interface. The routines for protecting and unprotecting
require kernel calls which leads to high performance penalties. Additionally, while
a page is unprotected other threads may still corrupt data. Furthermore, this
does not help against bit flips as they are not induced by stray writes, but by the
hardware itself. Furthermore, Bohannon et al. [4] handle the case for in-memory
database systems by computing XOR-checksums over certain protected memory
regions. A codeword table is maintained which stores the original checksums.
Pages are then later validated against this table. This again helps detecting
undesired, arbitrary writes, but bit flips may corrupt the codeword table and,
e.g., correct pages may then mistakenly be regarded as corrupted.

3   Error Detection by Arithmetic Codes

To tackle our vision of *resiliency-aware data compression techniques*, we decided to utilize arithmetic codes[2] as our resilience technique in a first step. Arithmetic codes are a long known technique to detect hardware errors at runtime caused by transient (e.g., dynamic bit flips) and permanent (e.g., stuck-at-1) hardware faults [23]. This is achieved by adding redundancy to processed data, i.e., a larger domain of possible data words is created. The domain of possible words contains the smaller subset of valid code words – the so-called encoded data items. Arithmetic codes are preserved by correct arithmetic operations, that is, a correctly executed operation taking valid code words as input produces a result that is also a valid code word.

3.1   Basic Idea of an Encoding

The underlying idea of AN encoding is simple: multiply each data word n by a predefined constant A, i.e., the code word $\hat{n}$ is computed as:

$$\hat{n} = n \cdot A \qquad (1)$$

As a result of this multiplication (*encoding*), the domain of values expands such that only the multiples of A become valid code words, and all other values are considered non-code. As an example, if one wants to encode a set of 2-bit numbers $\{0, 1, 2, 3\}$ with $A = 11$, then the set of code words is $\{0, 11, 22, 33\}$, while $1, 10, 34$ are all examples of non-code words.

If a bit flip affects an encoded value, the corrupted value becomes non-code with a probability of $(A-1)/A$. If $\hat{n} = 11$ and the least significant bit is flipped, then the new value $\hat{n}_{er} = 10$ and is non-code. To detect this fault, we have to check if the value is still a multiple of A:

$$\hat{n} \mod A = 0 \qquad (2)$$

Finally, to decode the value, we have to divide the code word $\hat{n}$ by A:

$$n = \hat{n}/A \qquad (3)$$

For any given native data width X – usually $X \in \{8, 16, 32, 64\}$ – processors' integer arithmetic modular arithmetic, i.e. the equation $a * b = c$ implicitly transforms into $|a*b| \equiv |c| \mod 2^X$ for unsigned integers, or $a*b \equiv c \mod 2^{X-1}$ for signed integers. By that, for several A's there exists a multiplicative inverse A^{-1} so that

$$n = \hat{n}/A = \hat{n} * A^{-1} \qquad (4)$$

For instance, $641^{-1} \equiv 6700417 \mod 2^{32}$ and Table 1 lists the available inverses for 32-bit unsigned integers for the given A's – in our case any even number has no inverse. Consequently, for some A's the division is replaced by a multiplication which is usually much faster on modern processors.

---

[2] Please note that some codes for lossless data compression are also called arithmetic codes. These are not equivalent with the ones used throughout this paper.

3.2   Beneficial Features of an Encoding

Generally, arithmetic code or AN-encoding offers some features that are beneficial for database systems. One of the features of AN encoding is the ability to directly process encoded data, i.e., there is no need to decode values before working on them. Most database-related operations can be performed on encoded values; these operations include addition, subtraction, negation, comparisons, etc. For example, the addition of two valid code words $11 + 22 = 33$ produces an expected code word, and 11 is less than 22 just like their original counterparts. Encoded multiplication and division are also possible, but require some adjustments.

This *encoded processing* feature is beneficial for in-memory database systems. AN-encoded data words can be read from main memory, processed using complex queries and stored back without the need for intermediate decoding, which reduces the overhead for resiliency mechanism. Examples of database operations on encoded data include scans, projections, aggregate computations, joins, etc.

3.3   Application Challenge

AN encoding is an arithmetic encoding scheme, allowing certain arithmetic operations directly on encoded data with relatively little overhead as well as multiplication and division with higher overhead. However, AN encoding does not pose any restrictions on a value of A. This constant must be carefully chosen to suit the needs of a particular application. The choice of A affects three parameters: *fault coverage*, *memory footprint*, and *encoding/decoding performance*. As a rule of thumb, greater values of A result in higher fault coverage, higher memory footprint and worse performance. The challenge is to find an A providing sufficiently high fault detection rate at a low cost of memory blow-up and performance slowdown.

In general, some "good" A's with the best trade-offs can be found. In terms of fault coverage, there is no known formula to find the best A, so the researchers resort to experimental results [11]. Memory blow-up depends on the size of A in bits; for example, encoding one 22-bit integer with a 10-bit A requires $22 + 10 = 32$ bits, i.e., an increase of 45%. Finally, performance slowdown can be negligible during encoding (since multiplication requires only $2 - 3$ CPU cycles, see the note on multiplicative inverses above), but can be a bottleneck during checks and decoding (since division is an expensive CPU instruction). To alleviate this decoding impact, A must be chosen such that the division operation is substituted by a sequence of shifts, adds, and multiplies [30].

4   Resiliency-Aware Data Compression

To the best of our knowledge, nowadays no additional information is added to explicitly detect bit flip corruption of compressed data in main memory database systems. In order to tackle an increasing bit flip error rate, in particular for dynamic errors, we want to tightly combine techniques from both fields of lightweight data compression and resilience techniques like AN encoding. On the one hand, lightweight data compression reduces or eliminates data redundancy to represent data using less bits. On the other hand, resilience techniques introduces data redundancy to detect bit flips. Therefore, both fields have opposing aims and combined approaches have to be carefully designed, so that benefits of both fields remain. Anyways, by compressing data, (almost) exclusively those bits which contain actual information are taken into account by the AN-encoding process. As we are going to show later, based on a well-defined and specific approach, the overhead of resiliency-aware data compression is less compared to uncompressed data, so that the approach is beneficial for database systems.

As next, we are going to present two specific AN-encoded compression scheme extensions: (i) AN-encoded Null Suppression in Sect. 4.1 and (ii) AN-encoded Run-Length Compression in Sect. 4.2.

4.1   AN-encoded Null Suppression

Null Suppression (NS) is the most well-studied kind of lightweight data compression technique. Its basic idea to the omission of leading zeros in small integers. This technique further distinguishes between bit-wise and byte-wise null suppression where either all leading zero bits or leading zero bytes containing only zero bits are stripped off. Usually, some kind of compression mask denotes how many bits or bytes were omitted from the original value. Decompression works by adding the leading zeros back.

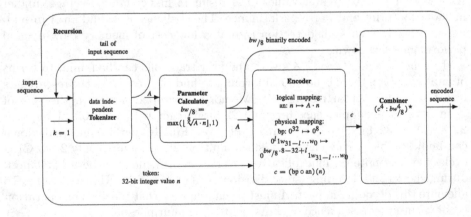

Fig. 1. Compression Scheme for Modularization for AN-encoded Null Suppression.

General Idea

The general idea of our AN-encoded Null Suppression technique is illustrated in Fig. 1. The illustration is based on our modularization concept for lightweight data compression methods [10]. Our scheme is a recursion module for subdividing data sequences several times. The first module in each recursion is a Tokenizer splitting the input sequence in finite subsequences or single values at the finest level of granularity. For that, the Tokenizer can be parameterized with a calculation rule. The finite output sequence of the Tokenizer serves as input for the Parameter Calculator, which is our second module. Parameters are often required for the encoding and decoding. Therefore, we introduce this module, whereas this module knows special rules (parameter definitions) for the calculation of several parameters. Our third module as depicted in Fig. 1 is the Encoder, which can be parameterized with a calculation rule for the processing of an atomic input value, whereas the output of the Parameter Calculator is an additional input. Its input is a token that cannot or shall not be subdivided anymore. In practice the Encoder gets a single integer value to be mapped into a binary code. The fourth and last module is the Combiner. It determines how to arrange the output of Encoder together with the output of the Parameter Calculator. Generally, these four main modules including the illustrated assembly in Fig. 1 are enough to specify a large number of lightweight data compression algorithms.

Our AN-encoded Null Suppression algorithm works as follows and is depicted in Fig. 1: We use a very simple Tokenizer outputting single integer values of an input data sequence. This Tokenizer instance can be characterized as *data independent* and *non-adaptive*, whereas only the beginning of the data sequence has to be known. For each value, the Parameter Calculator determines the number of necessary bytes (omission of leading zero bytes), whereas each value is multiplied by an value A for resiliency before. The corresponding formula is depicted in Fig. 1. The determined number of bytes is used in the subsequent Encoder to compute the bit representation of the AN-encoded value. The binary representations for whole compressed AN-encoded values are concatenated in the Combiner, symbolized by a star. That means, our AN-encoded Null Suppression technique encodes the values first and compresses afterwards. The compression mask itself is also AN-encoded.

SIMD-Based Implementation

In recent years, research in the field of lightweight data compression has mainly focussed on the efficient implementation of the techniques on modern hardware e.g., using vectorization capabilities of modern CPUs (SSE or AVX extensions). Schlegel et al. [24] presented 4-Wise Null Suppression as vectorized version. 4-Wise NS eliminates leading zeros at byte level and processes blocks of four integer values at a time. During compression the number of leading zero bytes of each of the four values is determined. This yields four 2-bit descriptors, which are combined to an 8-bit compression mask. The compression of the values is done by a SIMD byte permutation bringing the required lower bytes of the values together. This requires a permutation mask, which is looked up in an offline-created table using the compression mask as a key. After the permutation, the code words have a horizontal layout, i.e. code words of subsequent

```
basicstyle
1  compress (in elements[], out buffer[])
2  {
3    for (i = 0; i < |elements|; i = i + 4)
4    {
5      n1 = elements[i] * A;
6      ...
7      n4 = elements[i+3] * A;
8      z1 = count_zero_bytes(n1);
9      ...
10     z4 = count_zero_bytes(n4);
11     mask = (z4 << 6) | (z3 << 4) | (z2 << 2) | z1;
12     buffer ← (mask * A);
13     buffer ← n1;
14     ...
15     buffer ← n4;
16   }
17 }
```

Listing 1.1. Pseudo code for AN encoded 4-wise Null Suppression. elements is the input array while buffer is the output array. |elements| denotes the array's number of elements.

values are stored in subsequent memory locations. The compressed data is thus a sequence of compressed blocks. The decompression simply reads the compression mask, looks up the appropriate permutation mask which reinserts the leading zeros bytes and applies the permutation. Based on that principle, we are able to introduce our resiliency-aware extension of 4-Wise NS as an efficient vectorized implementation of our AN-encoded Null Suppression technique as illustrated in Fig. 1.

Encoding and Compression. Encoded compression for NS works as follows. Listing 1.1 shows the pseudo code for a 4-Wise encoded NS scheme (processing four 32-bit integers at once in a vectorized version). There are input and output arrays to function compress, where elements stores original data and buffer receives the compressed and encoded data. Four data items are processed in each loop iteration (line 3). First, each item is multiplied by A (lines 5–7) and afterwards the leading zero bytes are counted (lines 8–10). This can be done by counting the leading zero bits using compiler intrinsics (_builtin_clz() for g++) and then dividing by 8. The bit compression mask contains the number of leading zeros. It is computed by ORing the lower 2 bits of the zero byte counts together (line 11). Finally, the mask is encoded and the compressed encoded words are stored in the output buffer (lines 12–15). Assuming a little endian system, the leading zero bytes of a compressed value are inherently overwritten by the next appended data item, by advancing the write pointer by the number of non-zero bytes of the item just written.

```
 1 decompress (in buffer[], out elements[])
 2 {
 3   for (i = 0; i < |buffer|; )
 4   {
 5     mask = buffer[i] * A⁻¹;
 6     if (mask % A != 0) error();
 7     mask = mask * A⁻¹;
 8     i = i + 1;
 9     non_zero_bytes = mask & 0×3;
10     item = buffer[i] & (0xFFFFFFFF >> (non_zero_bytes * 8));
11     if (item % A != 0) error();
12     elements ← item * A⁻¹;
13     i = i + 4 − non_zero_bytes;
14     mask = mask >> 2;
15     non_zero_bytes = mask & 0×3;
16     ...
17   }
18 }
```

Listing 1.2. Pseudo code for AN encoded NS decompression. buffer is the input array while elements is the output array containing the uncompressed, decoded items. |buffer| denotes the array's number of elements.

Decompression and Decoding. Decompression and decoding is also straightforward. Listing 1.2 shows the according pseudo code. In this case, function decompress again receives an input and an output buffer and a loop iterates over the input buffer of AN-encoded and compressed data (lines 1,3). First, the compression mask is loaded, checked for errors and decoded (lines 5–7). Then, the buffer position is incremented (line 8), the number of non-zero bytes – denoted by the mask's lowest 2 bits – of the first data item is extracted (line 9) and the according bytes are stored (line 10). The restored item is checked against A and errors may be handled (line 11). Then, the decoded data item is stored in the output array and the read position of the input buffer is advanced by the number of non-zero bytes (lines 12,13). Finally, the mask is shifted right, so that the same steps can be repeated for the next three items, since always 4 items are represented by a single-byte compression mask (lines 14–16).

4.2 AN-encoded Run-Length Compression

The basic idea of Run-Length Compression (RLE) is to compress consecutive sequences of a same value – the *runs*. For compression, the distinct original value is stored together with the number of uninterrupted appearances – the *run length*. For decompression, these values are rolled out again.

General Idea

In contrast to our AN-encoded Null Suppression compression scheme, our AN-encoded RLE approach compresses first and encodes afterwards, since runs are

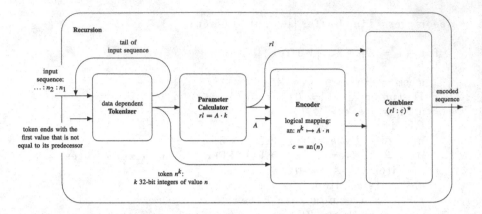

Fig. 2. Compression Scheme for Modularization for AN Run Length Encoding.

condensed to the value and its run length, therefore encoding only 2 values instead of long runs of values (see Fig. 2). That means, we reduce the necessary work for encoding using compression. In detail, AN-encoded RLE compression works as follows: Consecutive appearances of values are counted—the run lengths—using a *data dependent* `Tokenizer`. Whenever a new value is encountered, the previous value and its run length are encoded and written to the output. While the run-length is computed and encoded in `Parameter Calculator`, the value is encoded in the `Encoder`. The `Combiner` produces the resulting AN-encoded RLE output sequence. Decompression is done by reading in pairs of encoded values and their encoded run lengths. After checking both of them against A the decoded value is written "run length" times to the output buffer.

SIMD-Based Implementation

The SIMD variants differ only in comparing multiple input values against the current one, for compression, and in writing out multiple values at once. Since the encoding and decoding only takes place on the single values and their run lengths, changes to the algorithm are actually the same as to the sequential variant.

4.3   Summary

As shown in this section, the combination of AN-encoding and compression schemes differ, whereas the combination is straightforward. Nevertheless, the combination is useful from a database perspective and the AN-encoding integrates seamless in efficient vectorized compression techniques. However, our two examples are only a starting point and further research is necessary to protect compressed data in an efficient way. Additionally, the parameterization of the AN-encoding approach has a high impact as presented in the next section which is also a open topic.

Table 1. Information about A: parameter A; $invA$: the inverse for 32-bit integers (if applicable); $|A|$: the number of effective bits of A; $p_1 \ldots p_6$: the respective probabilities of not detecting $1 \ldots 6$ bit flips; *NS comp. rate*: NS compression rate for 16-effective-bits random integers; *overhead*: memory overhead of AN-encoded compression compared to the unencoded compressed ratio.

A	$inv(A)$	$|A|$	p_1	p_2	p_3	p_4	p_5	p_6	NS comp. rate	NS overhead
compr.		-	-	-	-	-	-	-	0.561	-
3	2,863,311,531	2	0.0	14.2	3.74	2.73	1.124	0.567	0.729	30 %
5	3,435,973,837	3	0.0	7.2	3.31	1.73	0.890	0.439	0.762	36 %
13	3,303,820,997	4	0.0	2.2	1.72	0.93	0.515	0.282	0.793	41 %
26	—	5	0.0	2.2	1.72	0.93	0.515	0.282	0.803	43 %
59	2,693,454,067	6	0.0	0.0	0.51	0.34	0.210	0.130	0.808	44 %
118	—	7	0.0	0.0	0.51	0.34	0.210	0.130	0.810	
250	—	8	0.0	0.0	0.25	0.19	0.133	0.088	0.811	45 %
507	2,837,897,523	9	0.0	0.0	0.08	0.07	0.046	0.040	0.936	67 %
641	6,700,417	10	0.0	0.0	0.08	0.06	0.040	0.030	0.962	71 %
965	485,131,021	10	0.0	0.0	0.00	0.04	0.032	0.025	0.996	78 %
7567	3,745,538,415	13	0.0	0.0	0.00	0.00	0.007	0.007	1.054	88 %
58659	2,839,442,059	16	0.0	0.0	0.00	0.00	0.000	0.001	1.061	89 %

5 Evaluation

In this section, we first discuss the choice of the constant A and what trade-offs it introduces. Then, we show the experimental results of applying AN encoding to Null Suppression and Run Length Encoding using 32-bit integers with only 16 effective bits to guarantee compressibility. We present throughput measurements for both sequential and SIMD implementations using $A = 641$. The experiments were run on a machine with an ASUS P9X79 Pro mainboard running a 12-core Intel i7-3960X CPU and 8×4 GiB (32 GiB) DRAM on an Ubuntu 15.04 OS. Generally, our measurements were executed on different sizes of random data sets, in particular 8, 16, 32, 64, 128, and 256 Million integers. Since all encoding / compression is done by copying instead of in-place operations, we use copying as a baseline. To ensure that the compiler does not generate undesired SIMD code, we use the GCC compiler flag `-fno-tree-vectorize`.

5.1 An Encoding

As mentioned earlier, the choice of parameter A affects the fault detection rate, memory blow-up, and encoding/decoding performance. Table 1 shows some

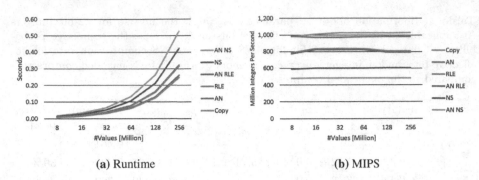

(a) Runtime   (b) MIPS

Fig. 3. Performance Evaluation of Sequential Algorithms.

"good" A's that range in their fault coverage[3], bit size and memory overhead, and whether there exists a multiplicative inverse for 32-bit arithmetic and thus fast decoding. For example, $A = 3$ has a size of 2 bits which leads to a 6% memory increase for 32-bit integers and we are able to detect all single bit flips but only 86% of double bit flips. On the other extreme, $A = 58,659$ ensures detecting up to 5 bit flips, but is 16 bits wide, leading to 50 % memory increase of 32-bit integers. In the end, the choice of A depends on how many bit flips should be detectable and how much redundancy is tolerable.

5.2   Compression Rates of AN-encoded Data Compression

AN-encoded Null Suppression. The last two columns of Table 1 show the typical compression rates for Null Suppression and the overhead of our AN encoding compared to pure NS. Notice that the original compression rate is about 0.561, and our AN-encoded NS scheme introduces a memory overhead of 30-89%. The overhead regarding uncompressed data reduces with smaller value ranges. For example, 16-bit compressible data using $A = 641$ occupies 3.8 % less space than uncompressed 32-bit data (0.962), while using ($A = 3$) occupies 27.1 % less space. The memory footprint of AN-encoded NS-compressed data exceeds uncompressed data when using an A which is more than 10 bits large.

AN-encoded Run Length Encoding. Since we assume 16 effective bits of data and $A = 641$ there is no actual memory overhead for RLE when comparing pure and AN-encoded RLE. If the values and run lengths are further compressed – e.g. using Null Suppression – then AN-coding incurs the overhead of the bits of the used A. RLE was tested with a fixed run length of 16.

---

[3] The probabilities for the table are taken from the experimental results of [11]; they can be found on https://www4.cs.fau.de/Research/CoRed/experiments.

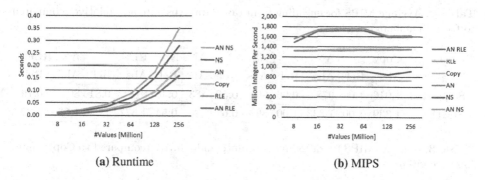

(a) Runtime

(b) MIPS

Fig. 4. Performance Evaluation of SIMD Algorithms.

5.3   Performance of Encoding / Compression

Figures 3 and 4 show the throughput measurements for our AN-encoded compression techniques for the sequential as well as SIMD (SSE 4.1) versions, respectively, using $A = 641$. In the experiments, we varied the size of the data sets. Table 2 lists the average number of MIPS, which is quit stable for all algorithms across the sizes of the sets.

Copying the data from one array to another is the baseline. AN-encoding itself leads to almost no overhead, which can be attributed to the good pipelining of simple multiplications. The SIMD RLE variant is faster than purely copying, because only an eighth of the original number of values is written: Instead of 16 32-bit integers, only 1 value and 1 run length, both 32 bits wide, are written.

For Null Suppression, AN-encoding incurs an overhead of $483/601 = 0.8020\%$ and $725/896 = 0.8119\%$ for sequential and SIMD variants, respectively. Next to the additional multiplication, the increased amount of data written to memory is responsible for the increase in runtime. The overhead for our AN-encoded RLE is $804/818 = 0.982\%$ and $1,639/1,677 = 0.982\%$ for sequential and SIMD variants, respectively. Depending on the run lengths, the multiplication for encoding is negligible, since much fewer ones than for NS are actually executed (one eighth, as described above).

5.4   Performance of Decoding / Decompression

Table 3 outlines the average MIPS for decoding / decompressing from the encoded / compressed formats. As expected, pure AN-coding exhibits (almost) the same results as encoding, since there exists a multiplicative inverse for $A = 641$ and the differences are negligible.

Decoding Null Suppression is slower than Compression, since more data is written back to memory than during decoding. AN-encoded NS decoding becomes much worse now, because every encoded value is first tested against A using the modulo operation and then decoded back by multiplying with the

Table 2. Average MIPS for encoding from raw data. Absolute and relative (compared to Copy) numbers are given.

	Copy		AN		NS		AN+NS		RLE		AN+RLE	
	Abs	Rel	Abs	Rel	Abs	Rel	Abs	Rel	Abs	Rel	Abs	Rel
Sequential	1,016	1.00	978	0.96	601	0.59	483	0.48	818	0.81	804	0.79
SIMD	1,339	1.00	1,333	1.00	896	0.67	725	0.54	1,677	1.25	1,639	1.22

Table 3. Average MIPS for decoding. Absolute and relative (compared to Copy) numbers are given.

	Copy		AN		NS		AN+NS		RLE		AN+RLE	
	Abs	Rel	Abs	Rel	Abs	Rel	Abs	Rel	Abs	Rel	Abs	Rel
Sequential	1,016	1.00	1,013	1.00	500	0.49	260	0.26	1,051	1.03	994	0.98
SIMD	1,339	1.00	1,317	0.98	750	0.56	570	0.43	1,258	0.94	1,339	1.00

inverse. Decoding overead is as high as $260/500 = 0.5248\%$ and $570/750 = 0.7624\%$ for sequential and SIMD code, respectively.

RLE decoding is very fast since on the one hand much less memory is read compared to what is read – due to the run length of 16 – and on the other hand unrolling of values is much simpler than NS decoding.

5.5   Summary

We find these results encouraging. The choice of parameter A provides trade-offs in terms of error detection capability, memory overhead, and performance penalty, while the speeds of the compression schemes are affected differently by the encoding overhead – both in terms of added complexity of code as well as dependency on the data characteristics.

6   Conclusion

Modern database systems are very often in the position to store their entire data in main memory. The reasons are manifold: (i) increased main memory capacities, (ii) column-oriented storage format and (iii) lightweight data compression techniques. Unfortunately, hardware becomes more and more vulnerable to random faults, so that e.g., the probability rate for bit flips in main memory increases, and this rate is likely to escalate in future dynamic random-access memory (DRAM) modules. Since the data is highly compressed by the lightweight compression algorithms, multi bit flips will have an extreme impact on the reliability of database systems. To overcome this issue, we have introduce our research on error resilient lightweight data compression algorithms in this paper. In detail, we have utilized arithmetic AN encoding, which is one family of codes which is an interesting candidate for effective software-based error detection. We have presented two algorithms: (i) AN-encoded Null Suppression and (ii) AN-encoded Run Length Encoding. We have shown in our experiments that by using AN encoding, much higher bit flip detection capabilities are achievable than with SECDED ECC. Furthermore, Our evaluation indicates that data compression schemes augmented with AN encoding become resilient at a low memory and performance cost. As an example, a "golden" A of 641 makes 16-bit data completely resilient to single and double bit flips. Depending on the scenario and the compression schemes, AN encoding results in little to no performance penalties. Of course, encoding leads to memory overhead, but we also showed that gains over uncompressed data are still possible. For instance, AN encoded Null Suppression occupies 4% less space than uncompressed data, with a 5-10% slowdown of compression/decompression speed for the case of SIMD.

Acknowledgements. This work is partly supported by the German Research Foundation (DFG) within the Cluster of Excellence "Center for Advanced Electronics Dresden" (cfAED) and by the DFG-grant LE-1416/26-1.

References

1. Abadi, D., Madden, S., Ferreira, M.: Integrating compression and execution in column-oriented database systems. In: Proceedings of the 2006 ACM SIGMOD International Conference on Management of Data, pp. 671–682 (2006)
2. Antoshenkov, G., Lomet, D.B., Murray, J.: Order preserving compression. In: Proceedings of the Twelfth International Conference on Data Engineering, ICDE 1996, pp. 655–663 (1996)
3. Bassiouni, M.A.: Data compression in scientific and statistical databases. IEEE Trans. Softw. Eng. **11**(10), 1047–1058 (1985)
4. Bohannon, P., Rastogi, R., Seshadri, S., Silberschatz, A., Sudarshan, S.: Detection and recovery techniques for database corruption. IEEE Trans. Knowl. Data Eng. **15**(5), 1120–1136 (2003)
5. Boncz, P.A., Manegold, S., Kersten, M.L.: Database architecture optimized for the new bottleneck: memory access. In: Proceedings of the 25th International Conference on Very Large Data Bases, VLDB 1999, pp. 54–65 (1999)
6. Chen, Z., Gehrke, J., Korn, F.: Query optimization in compressed database systems. SIGMOD Rec. **30**(2), 271–282 (2001)
7. Goldstein, J., Ramakrishnan, R., Shaft, U.: Compressing relations and indexes. In: Proceedings of 14th International Conference on Data Engineering, pp. 370–379, February 1998
8. Graefe, G., Kuno, H., Seeger, B.: Self-diagnosing and self-healing indexes. In: DBTest, pp. 8:1–8:8 (2012)
9. Graefe, G., Stonecipher, R.: Efficient verification of b-tree integrity. In: BTW, pp. 27–46 (2009)
10. Hildebrandt, J., Habich, D., Damme, P., Lehner, W.: Modularization of lightweight data compression algorithms. Technical report, Department of Computer Science, Technische Universität Dresden, November 2015. https://wwwdb.inf.tu-dresden.de/misc/team/habich/dcc2016.pdf. submitted to DCC 2016

11. Hoffmann, M., Ulbrich, P., Dietrich, C., Schirmeier, H., Lohmann, D., Schröder-Preikschat, W.: A practitioner's guide to software-based soft-error mitigation using AN-codes. In: HASE 2014, pp. 33–40 (2014)
12. Huffman, D.A.: A method for the construction of minimum-redundancy codes. Proc. Inst. Radio Eng. **40**(9), 1098–1101 (1952)
13. Hwang, A.A., Stefanovici, I.A., Schroeder, B.: Cosmic rays don't strike twice: understanding the nature of DRAM errors and the implications for system design. SIGARCH Comput. Archit. News **40**(1), 111–122 (2012)
14. Kissinger, T., Kiefer, T., Schlegel, B., Habich, D., Molka, D., Lehner, W.: ERIS: a numa-aware in-memory storage engine for analytical workload. In: International Workshop on Accelerating Data Management Systems Using Modern Processor and Storage Architectures - ADMS, pp. 74–85 (2014)
15. Kolditz, T., Kissinger, T., Schlegel, B., Habich, D., Lehner, W.: Online bit flip detection for in-memory b-trees on unreliable hardware. In: DaMoN, pp. 5:1–5:9 (2014)
16. Lehman, T.J., Carey, M.J.: Query processing in main memory database management systems. In: Proceedings of the 1986 ACM SIGMOD International Conference on Management of Data, SIGMOD 1986, pp. 239–250 (1986)
17. Lehner, W.: Energy-efficient in-memory database computing. In: Design, Automation and Test in Europe, DATE 13, Grenoble, France, 18–22 March 2013, pp. 470–474 (2013)
18. Lemire, D., Boytsov, L.: Decoding billions of integers per second through vectorization. CoRR abs/1209.2137 (2012)
19. May, T.C., Woods, M.H.: Alpha-particle-induced soft errors in dynamic memories. IEEE Trans. Electron Devices **26**(1), 2–9 (1979)
20. Moon, T.K.: Error Correction Coding: Mathematical Methods and Algorithms. Wiley, Hoboken (2005)
21. Reghbati, H.K.: An overview of data compression techniques. IEEE Comput. **14**(4), 71–75 (1981)
22. Roth, M.A., Van Horn, S.J.: Database compression. SIGMOD Rec. **22**(3), 31–39 (1993)
23. Schiffel, U.: Hardware Error Detection Using AN-Codes. Ph.D. thesis, Technische Universität Dresden (2011)
24. Schlegel, B., Gemulla, R., Lehner, W.: Fast integer compression using simd instructions. In: DaMoN. pp. 34–40 (2010)
25. Schroeder, B., Gibson, G.A.: A large-scale study of failures in high performance-computing systems. Dependable Secure Comput. **7**(4), 337–350 (2010)
26. Schroeder, B., Pinheiro, E., Weber, W.D.: Dram errors in the wild: a large-scale field study. In: Proceedings of the Eleventh International Joint Conference on Measurement and Modeling of Computer Systems, SIGMETRICS 2009, pp. 193–204 (2009)
27. Stepanov, A.A., Gangolli, A.R., Rose, D.E., Ernst, R.J., Oberoi, P.S.: Simd-based decoding of posting lists. In: Proceedings of the 20th ACM International Conference on Information and Knowledge Management, CIKM 2011, pp. 317–326 (2011)
28. Stonebraker, M.: Technical perspective - one size fits all: an idea whose time has come and gone. Commun. ACM **51**(12), 76 (2008)
29. Sullivan, M., Stonebraker, M.: Using write protected data structures to improve software fault tolerance in highly available database management systems. In: VLDB, pp. 171–180 (1991)
30. Warren, H.S.: Hacker's Delight. Addison-Wesley Longman Publishing Co., Inc., Boston, MA, USA (2002)

31. Witten, I.H., Neal, R.M., Cleary, J.G.: Arithmetic coding for data compression. Commun. ACM **30**(6), 520–540 (1987)
32. Ziv, J., Lempel, A.: A universal algorithm for sequential data compression. IEEE Trans. Inf. Theor. **23**(3), 337–343 (1977)
33. Zukowski, M., Heman, S., Nes, N., Boncz, P.: Super-scalar ram-cpu cache compression. In: Proceedings of the 22nd International Conference on Data Engineering, ICDE 2006, pp. 59–59, April 2006

Author Index

Printed in the United States
By Bookmasters